A TABLE FOR ONE

MANCHESTER
1824

Manchester University Press

A TABLE FOR ONE

A critical reading of singlehood, gender and time

Kinneret Lahad

Manchester University Press

Published by Manchester University Press
Altrincham Street, Manchester M1 7JA

www.manchesteruniversitypress.co.uk

British Library Cataloguing-in-Publication Data
A catalogue record for this book is available from the British Library

ISBN   978 1 5261 1539 3   hardback
ISBN   978 1 5261 1635 2   open access
ISBN   978 1 5261 1727 4   paperback

First published by Manchester University Press in hardback 2017.
This edition first published 2019.

Typeset
by Toppan Best-set Premedia Limited

Contents

Acknowledgements

The publication of this book bears the imprint of many hands. First, I wish to thank my mentor Hanna Herzog for being an outstanding, kind and inspiring person in my life. Thank you for being my brilliant academic mother, for your warm encouragement, profound humanity, and for allowing me to learn alternative modes of doing feminist scholarship. Zeev and you have been true friends and role models. I am deeply indebted to Haim Hazan for our priceless intellectual exchanges, his unwavering support and belief in my scholarship along the way. Our meetings have always been an endless source of pertinent questions and theoretical sophistication. Thank you for your creative ideas, for your bright and unexpected inputs and for formulating together some of the analytical threads of this project. I cannot thank both of you enough.

I also thank Ilana Silber for her support, sharing with me her knowledge of sociological theory and for being one of the first scholars who made me realize how much I love sociology. My sincere thanks to Eviatar Zreubavel: his work on social time has provided a benchmark for my thoughts on temporality. I extend sincere thanks to Hanna Naveh, whose work on family life has sparked my interest in researching family and the mysterious organization of personal life. Avenr De-Shalit, Zeev Sternhal, Carola Hilfrich, Orly Lubin, Louise Bethlehem, and Hannan Hever provided me with critical guidance many years ago and opened my eyes to understanding the richness and beauty of textual analysis.

My deepest thanks to my close and supportive circle of friends. I am indebted to Tal Kohavi for our special friendship and for her intellectual guidance, inspiration, warmth, and generosity. I particularly want to thank Karmit Haber, who is a rock of support and strength in my life, and Michal-Kravel Tovi, for her friendship, good humor and excellent advice and suggestions. I also want to thank Anat Hammermann Schuldiner and Lina Yassin for their affection and assistance in more ways than one. Special thanks to Orna Donath and Avi Shooshna for influencing my ideas and for their keen interest and enthusiasm at various stages of this project. I also want to express my appreciation to Carolin Aronis, Yaniv El-Ron, and Adi Moreno for offering thoughtful feedback on various versions of the chapters and for their unwavering support. I extend my thanks to Dafna Hirsch, Smadar Sharon and Omer Sharon Gabay for their care, love, and encouragement. During the last phases of the book I was very fortunate to meet my brilliant friend Irit Dekel, whose support and encouragement are exemplary: thanks to you and Michael Weinman for inviting me to your home.

At Tel-Aviv University I have been fortunate to work with exceptionally wonderful colleagues. A very special thanks to Daphna Hacker for her support and for being a model of personal scholarly integrity and feminist veracity. I am grateful as well to Ofra Goldstein-Gidoni for her friendship, faith, and many brainstorming sessions over good coffee and cake. For her goodwill, encouragement, and significant support I thank Smadar Shiffman, and I am sincerely grateful to Michèle Bokobza Kahan and Yofi Tirosh for their support and encouragement. I extend my warm thanks to my intellectual soulmate and friend Vanessa May. My warm thanks and appreciation go to Karen Hvidtfeldt Madsen for her special friendship, interest, and support of this project. I also want to express my enduring gratitude to Gokce Yurdakul for her unflagging support. Special thanks for many stimulating conversations with my wonderful friends and colleagues Yvette Taylor, Charlotte Krol økke, and Sarah Wilson. My heartfelt thanks to all of you.

During my travels in and outside Israel I was lucky to meet such inspiring colleagues and friends as Niza Yanai, Lynn Jamieson, Rachel Reidner, Shelley Budgeon, Khalad Furani, Yehouda Shenhav, Aeyal Gross, Iris Rachmaniov, Yael Hashiloni-Dolev, Tami Herzig, Inna Leykin, Pnina Lahav, Niza Berkovitz, Miri Eliav Feldon, Christiane Brosius, Jeroen De Kloet and Laila Abu-Er-Rub. I wish to thank my colleagues and friends at Venice International University and Nili Cohen for providing me with a beautiful office and a stimulating environment to write my book. I thank Gil Eyal and Ynon Cohen from Columbia University for their hospitality. More thanks to Hedva Abu Hassira Marsh, Anat Lapidot Firilla, Zippi Hecht, Charlotte Whiting, Cecile Moskovitz, Nicolas Seutin, Tamsin Sanderson, Tea Stifanic, Siobhan Kattago, Sara Armstrong, and Lucy Pickering.

My deep appreciation to Akin Ajayi, Maayan Shtendel, Neta Yodovich, and Yael Braudo-Bahat for their excellent editing, patience, diligence, and professionalism. I have learnt so much from you and I am sincerely grateful. It has been a great privilege to work with the staff of Manchester University Press: foremost thanks to Thomas Dark for his enthusiasm, belief, and support of this book, and to Robert Byron, Alun Richards, and Danielle Shepherd for shepherding me through the publication process. I also wish to express my gratitude to the anonymous reviewers, who provided insightful suggestions which have improved the final version of the manuscript.

Thanks to the institutions that have funded my research and given me time to write. I received significant support from the Department of Sociology and Anthropology in Bar Ilan, the Department of Sociology and Anthropology at Tel-Aviv, the *Porter Institute* for Poetics and Semiotics, and the NCJW women and gender program. This research was also supported by the *Israel Science Foundation* (grant 8/17). My heartfelt appreciation goes to Orly Lubin, Jenny Barak, Lisa Ben-Senior, Michal Shoef, Seffi Shtiglish, and Ariel Pridan. I owe gratitude to Qingling Guo for allowing me to use her beautiful artwork as the book cover, and Penn Ip for her kindness, significant help, and translation from Chinese to English. Earlier and different versions of Chapters 4 and 7 have appeared in *Sociological Forum* and *Women's Studies International Forum*: I thank the editors for their kind permission to reprint them here.

Lastly, very special thanks to my family, who have endured the writing of the book at close range: I love you dearly. To my parents Judy and Gad Lahad for their energy, curiosity, and wild humor. Special thanks to my mom for instilling me with a love for travel, the value of female companionship, and an appreciation of literature and culture. My dad has taught me the value of being dedicated to something you love and has been a role model of goodness and candidness. This book was written in loving memory of my grandmothers and grandfathers, Menachem and Mirian Peled, Elisheva, and Yitzhak Lahad.

I would like to extend special thanks to my kind and loving brother Yarden Lahad, whose faith in me is really priceless, and to Yael Lahad Gelfestein for their warm, deep care, and generous hospitality in my frequent visits in Berlin. Many thanks to my beloved sister Maayan Shanan for being in my life; I thank her and her partner Omri Shanan for their support and providing me with many hours of pleasant distractions. I thank my cousins and best friends Shirli Brosh, Tali Raveh and Gili Raveh, and their parents, Yehuda and Tami Raveh: heartfelt thanks for your love and encouragement. To my aunts Hani Ben-Ari and Malka Biron, Yoav, Neri, Amnon, and Ephrat Levi-Lahad and also in memory of Miriam and Shlomo Dinur, all of you have given me support to dream and write. Lastly, I thank my dear and beloved nieces and nephew Neta, Nitzan, Noam, Niri, and Yuval—always asking the most inquisitive questions and making my life so rich, funny, and colorful. I dedicate this work to you.

1

Introduction

A few years ago a dear colleague of mine asked how it could be that a woman like me was *still single*. She simply couldn't "figure me out," she said; her facial expression was one of genuine concern. To her, it seemed that I was neither actively looking for a male partner nor concerned by my overly extended singlehood. As a thirty-plus single woman at the time, I had become accustomed to this line of speculative questioning, one which expected me to justify my personal social circumstances. This time, however, rather than parrying the question, I decided to take a different route and turned the question back on her.

Rather than trying to justify my single existence—or, alternatively, refusing to answer to what I've often considered to be an intrusive and a non-dialogical form of interrogation—I asked her if she was happily married. I did have the advantage of prior knowledge, it should be said; my question arose from previous conversations, in which my colleague had freely discussed her martial difficulties. Because we were habitually frank with one another, I knew that she was unhappily married, and that she and her husband had been living separate lives for some years. With this knowledge in mind, I asked her what stopped her from separating from her husband. "It's complicated," she replied. "Well, single life can be quite complicated too," I retorted.

A recurring experience for many single women is the need to account for their singlehood. This demand is articulated in different ways—blatant and subtle, explicit and implicit—but seems to lead to the same end. *Why are you still single? What's wrong with you? Why aren't you trying hard enough? Shouldn't you lower your standards?* This set of statements—part speculative, part unprompted advice, at times a blend of pity, alarm, and scorn—is a constant scourge for many single women. The forms of interactions in which single women are constantly required to account for their status are ubiquitous. It happens during family dinners, encounters with friends (and their friends), conversations with neighbors and even interactions with total strangers. Researching female singlehood for almost a decade has shown that many people are perplexed by what they consider to be a disturbing enigma: a woman in her "prime years," who has not entered matrimony and is yet to embrace the familial way of life. And so, otherwise banal questions have become the anchor for popular romantic comedies, matchmaking reality shows, catchy newspaper article headlines, and the titles of bestselling self-help books worldwide.

Scholars and social commentators have extensively analyzed the growing population of single women worldwide, in an attempt both to understand the phenomenon and to propose alternatives to the popular portrayal of female singlehood. In 2014, the percentage of single-person households in England and Wales was 28.4%; in Israel, in 2013, it was 18.7%. In 2014, single women were 30.7% of the female population of England and Wales (6.45 million; Office for National Statistics 2015), while in Israel in 2013, they were 28.1% of the female population (830 thousand; Central Bureau of Statistics 2015), compared to the 54.6% who were married.[1] Similar percentages of single women have also been seen in other countries in recent years, such as the United States and Denmark, where singles are also almost 30% of the female population (40 million and 850 thousand, respectively; United States Census Bureau 2016; Statistics Denmark 2016).

This significant demographic shift has produced new conceptualizations of singlehood: "leftover women" in China; "parasite women" in Japan; "late singlehood" in Israel; the "singletons" of Australia, the US, and the UK. Whatever the name, the intent is to capture both the cause and the effect of singlehood on society. What is clear is that the phenomenon stirs public debate on a global scale—a debate that considers extended female singlehood as both a disturbing and an exciting new phenomenon.

Drawing on a wide range of cultural resources—including web columns, blogs, advice columns, popular clichés, advertisements, and references from television and cinema—I will attempt to outline some of the meaning-making processes of singlehood and time in Israel. The case study of Israel carries broader implications for contemporary discussions about singlehood and time in general, because it presents, in particular, the opportunity to raise questions about processes of continuity and change, transition, and tradition. These are all highly relevant concepts for societies undergoing dramatic shifts in personal relationships and the way new forms of femininity are regulated (Budgeon 2015).

It should be highlighted that the academic literature on singlehood often tends to group together different forms of non-marriage. However, singlehood is not a homogenous category of membership and social relations. Indeed, widowhood, divorce, and single parenthood are sometimes all conceptualized under the general umbrella of singlehood. Although there are undoubtedly many shared discursive patterns binding these categories, nonetheless some of the fundamental disparities between them are regularly overlooked. My working definition of singlehood throughout this study refers to long-term singles, whom according to socially constructed parameters are considered as "aging single women" and are ascribed with the category of "late singlehood." These are women who are not in a committed long-term relationship, and do not have children. I also include divorced women, who did not remarry, nor subsequently feel the inclination to remarry. It is important to stress that I do not include in this research the social categories of single mothers, or widows, and neither do I include women who share their lives with a permanent partner.

The term "late singlehood" has evolved in the Israeli public discourse since around the year 2000. The term designates men and women whose single status is no longer regarded as socially acceptable, and mirrors terms used elsewhere, such as "always single" (Maeda and Hecht 2012) or "extended singlehood" (Sharp and Ganong 2011). By referring to the term I will specifically deal with representations of the socio-temporal phase wherein singlehood becomes a "problem in the eyes of society," or, in other words, when it ceases to be considered as a "normative stage" preceding marriage and parenthood.

I have decided to use the term "late singlehood" throughout this study as it emerges from the Jewish Israeli public discourse about single persons, across both religious and secular populations. It is interesting to note how the term has also been adopted in both lay discourse as well as the language of psychological and popular therapeutic treatment offering "cures" for late singlehood. By using the term "late singlehood," I aim to reflect and to be attuned to contemporary discourses which reveal how "late singlehood" and "aging single women" defy the hegemonic temporal norms, thus creating what Diane Negra describes as a "feminized temporal crisis" (Negra 2009, 54). My definition of "late singlehood" in this study (ranging from the mid–late twenties onwards) is somewhat arbitrary but non-arbitrary at the same time. My contention is that late singlehood is a non-scheduled, non-institutionalized transition process; therefore it does not entail precise entry or exit passages which can be defined, mandated, or celebrated. Nonetheless, although these processes are individualized, they are socially synchronized with collective schedules.

With this in mind, I stress that singlehood is a contingent notion which varies according to gender, age, class, religion, ethnicity, ableness, sexual orientation, or other axes of social differentiation. This definition, then, takes into account a feminist inter-sectional methodological approach as it recognizes that singlehood is not a homogenized category of one identity and is formed through different positionalities and distinctive structures of oppression (Collins 2000).[2]

Thus, this analysis also seeks to contribute towards conceptualizing a more complex gender intersectional analysis, by adding singlism (DePaulo 2006; DePaulo and Morris 2005)—namely the discrimination and mistreatment of single persons—as a category for analysis. In other words, there is thus a necessity for future studies to incorporate singlehood and relationship status in feminist theorizing of intersectionality. Therefore, my interest here is in considering how patriarchy and heteronormativity overlap and intersect with other structures of domination such as singlism and ageism, and are carried out through gendered configurations of time. It is important to stress that singlism is a socially shared belief, one that impacts upon multiple facets of life including housing, wages, and unequal access to services and benefits (DePaulo 2006). As DePaulo (2006) and Hacker (2001) point out, single persons are often excluded from discounted health benefits, greater social security options, lower tax bills, and higher salaries. This point will be further developed in the last chapter of this study.

To put it another way, the experience of singlehood intersects with various factors which manifest themselves in varying social contexts and are therefore subject to different forms of exclusions, privileges, and discrimination. Moreover, Israel is a society characterized by various cleavages, such as the Jewish-Arab cleavage, the secular-religious cleavage, the ethnic cleavage, and class cleavage (Ben-Porat 2006; Horowitz and Lissak 1989). Thus, when I refer here to the concept of singlehood I do not include "all" single women, but rather refer mainly to heterosexual, cisgender, white, Jewish (in the Israeli context), middle-upper-class, able-bodied single women. Given these parameters, I do not analyze experiences of women who identify themselves as religious, as lesbians, or as having a disability, for example. In that respect, this book is also a call for future studies that can examine the nuanced influences of religious beliefs, homophobia, or ableism upon the experience of singlehood.[3]

My choice to focus on single women and not single men derives from my attempt to understand how patriarchy and heteronormativity affects women's lives. From this vantage point, I seek to re-examine the effects of what Adrienne Rich (1980), in her seminal article, termed *compulsory heterosexuality*. By this, I refer to the cultural, social, financial, and other mechanisms that direct women into being sexually involved with men and deny the possibility of sexual, as well as emotional, intimacy with other women. Or, in other words, the social forces or structures which maintain women as sexually, emotionally, and reproductively available to men. This is one of the reasons why long-term singlehood is still seen by many as not representing a viable option for women, because it does not conform to gendered expectations and defies gender socialization in general. Moreover, the fact that long-term singlehood is not perceived as a feasible possibility may be one reason why some women remain in unhappy and even abusive relationships. In this manner, my work corresponds with feminist criticism which has long sought to debunk the traditional discourse of feminine ideals. For example, I want to challenge the assumption that the status and social worth of women is dependent upon and defined in terms of their relationships to men, or the prevailing conviction that the primary role of a woman is to care for her family members.

Putting singlehood on the critical desktop

An inspiring number of critical works about single women have been published since 2000.[4] One key research direction underlying these works is the attempt to scrutinize and debunk the widespread myths and stereotypes attached to single women. To a large extent, these studies confirm the impression that single women in different parts of the world are regularly typecast as desperate, hysterical, childish, irresponsible, or lazy.[5]

Certainly, these studies have succeeded in promoting the voices and experiences of single women, as well as introducing a more diversified picture of their everyday lives. This book joins these significant endeavors in establishing singlehood as a field of study that warrants separate consideration on its own terms (Byrne 2009). As such, it stresses that the study of singlehood should take into consideration not just the prevalence of these ideologies, but also the need to direct its attention to the paradigms and strict categorization which constantly define and limit what singlehood means and stands for. Undoubtedly, this is a significant challenge. My claim is that in order to understand what infuses these pejorative interpretations of female singlehood with such discursive force, we need to deepen our understandings of their sources, social mechanisms, and consequences.

The groundwork for understanding the social meaning of singlehood, I argue, can be drawn from one of the central arguments of this book: the concept and comprehension of *Time* plays a crucial role in the discursive formation of traditional conventions about female singlehood, and in the production of single women's subjectivities. It is a premise of my study that our understanding of singlehood is dominated by unquestioned temporal models, premises, and concepts. This is one reason why one cannot understand these everyday dynamics, as well as the natural and authoritative tone through which they are conveyed, without understanding how over-determined frameworks of temporal categories are constituted. This might also be why it can be so difficult to resist and challenge many of the convictions about singlehood, because they are articulated through the language of time, a language characterized by its normative self-evident positions and regimes of truth.

This study seeks to locate singlehood within a broader critical theory and context. To achieve this goal, I juxtapose two theoretical subfields that are rarely linked: the social study of Time, and the study of Singlehood.[6] This conjunction of two supposedly separate bodies of knowledge can be of benefit to one another. For one thing, temporality plays a crucial role in the formation of singlehood; at the same time, analyzing singlehood can shed fresh light on how temporal orders are constructed and maintained. Indeed, this integration demands the rethinking and the reconfiguration of the categories and cultural forces that create the framework through which singlehood and temporal orders are constituted.

The socially related studies of Time can offer us both a new analytical framework and the innovative conceptual vocabulary from which we can reassess some of our dominant taken-for-granted conceptual frameworks. They give us the opportunity to explore and theorize singlehood through temporal concepts such as *Ageism and accelerated aging* (Chapter 4), *Temporal economy* (Chapter 5), and *Waiting* (Chapter 7). Other temporal categories which are examined throughout this book, such as age, the life course, linearity, and heteronormativity, enable a fresh consideration of our dominant perceptions about collective clocks, schedules, time tables, and the temporal organization of social life in general. By proposing this new analytical direction, this book seeks to rework some of our common conceptions of singlehood, and presents a new theoretical arsenal with which the temporal paradigms that devalue and marginalize single women can be reinterpreted.

We give little thought to the everyday workings of socio-temporal templates, and how these underpin most of our thought habits, social practices, and everyday interactions. However, the interpretation of these socio-temporal constructs—as I will attempt to show in the subsequent chapters—will reconfigure our understanding of the ways by which temporal knowledge is constructed, and will question the very terms upon which this is based. In this context, central to my study is the fundamental sociological question about how meaning is produced, as well as how temporal assumptions about singlehood are consolidated through interactions with others. From this perspective, neither singlehood nor time can be fixed and neutral categories, as they are constituted through changing social contexts, discourses, and human interactions. In this vein, I import some of the basic ideas advanced by discourse analysis, social constructionists, and symbolic interaction approaches, as well as ideas taken from feminist and queer scholarship.

The surge in singlehood literature published from around 2000 onwards undoubtedly contributes to a more critical reading of prevalent representations of single women. It also challenges widespread hegemonic assumptions about them. The literature responds to what is now a well-established fact, namely that more and more people are living on their own. Scholars like Shelly Budgeon (2008; 2015) Michael Cobb (2012), Bella DePaulo (2006), Lyn Jamieson and Roona Simpson (2013), Lyne Nakano (2011, 2014), Jill Reynolds (2008), Jesook Song (2014), and Anthea Taylor (2012) point out that despite the global growth of single-person households, late singlehood is still commonly perceived in terms of negations: a lack, an absence, a deficient identity. Questions like "Why are you single?" and "What is wrong with you?" (as my colleague asked me), and the dominant image of single women as lonely, desperate "cat ladies" embody this view.

Underpinning these attitudes is also a fear of female singlehood, which has been allowed to exceed its temporal boundaries. For many, long-term singlehood represents a threat to social order, and to subjectivity, thus demanding increased scrutiny and control. Central to this perspective is the assumption that singlehood can only be a temporal, liminal transitory status, during which single women can only hope to *unsingle* (DePaulo 2008) themselves and get married.

Consequently, long-term singlehood cannot possibly be a desired or chosen position. Implicit in this is the assumption that singlehood, when chosen at a younger age, will slowly and inevitably degrade into the miserable, vulnerable, lonely life of an "aging old maid." As I have argued elsewhere (Lahad 2013; 2014), chosen spinsterhood, or the notion of an "old maid" by choice appears to be a contradiction in terms, as though no one could possibly wish to grow older as a single woman. These well-worn stereotypes have a powerful presence in popular imagery, associating the category of the old maid with an unfortunate sequence of events and an empty and lonely form of existence.

The new scholarship on singlehood shows that there are increasing numbers of single women who report high levels of life satisfaction, and others who refuse to compromise and marry men who have not met their marriage expectations.[7] Singlehood is not, by any stretch of the imagination, automatically a catastrophe. Eric Klinenberg's (2012) *Going Solo: The Extraordinary Rise and Surprising Appeal of Living Alone*, which received significant media exposure in the US, dismisses the widespread assumption that living alone necessary leads to isolation, misery, and loneliness. He notes that many single persons enthusiastically embrace singular forms of living, and are content with their single status.

However, being single does not always necessarily imply living alone. First, it is worth noting that the capacity for living alone may depend upon one's material resources. Second, a different perspective on singlehood and prevailing living arrangements may also take into consideration more varied household compositions, such as co-housing and community housing, which do not necessarily subscribe to the conception of the nuclear family household unit. These issues are beyond the scope of the current discussion, but is it important to note that these alternative living arrangements can promote other forms of economic and emotional exchanges, and encourage a more ecological and environmentally friendly mindset.

Further studies also disclose diverse responses to solo living. Interviews with single women conducted by scholars like Tuula Gordon (1994), Jill Reynolds (2008), and Lyne Nakano (2011) reveal that some of their respondents fluctuate between choosing and non-choosing singlehood, or occupy the subject position of singlehood by chance. And yet, despite what appears to be a dramatic demographic shift, it seems that the stereotypes and mythical narratives of single women as desperate, lonely, and miserable remain as prevalent as ever.

My approach to understanding singlehood is very much influenced by Haim Hazan's (2002) approach to the study of old age. Hazan suggests that the aged should be seen as *carriers of the cultural tag of old age* (ibid., 232). Based upon this theoretical formulation, I would like to suggest that conceptualizing single women as *carriers of the cultural tag of singlehood* can illuminate more discursive dimensions and open up new avenues for the analysis of social life.

Another important source of inspiration is the work of Rosemarie Garland-Thomson (2002), a feminist disability scholar. In a study calling for the integration of feminist and disability theory, Garland-Thomson claims that:

> There has been no archive, no template for understanding disability as a category of analysis and knowledge, as a cultural trope, and an historical community. So just as the now widely recognized centrality of gender and race analyses to all knowledge was unthinkable thirty years ago, disability is still not an icon on many critical desktop. (ibid., 2)

I would like to make a similar claim with regard to singlehood. Singlehood has no archive, and does not act as a category of analysis and knowledge. Borrowing Garland-Thomson's formulation, I argue that singlehood lends a new perspective to critical theory and possesses the potential to enrich sociological, feminist, disability, and queer theory.

Queer theory provides a significant conceptual lens to this study. In his reassessment of queer politics, Michael Warner contends that many of the "environments in which lesbian and gay politics arises have not been adequately theorized and continue to act as unrecognized constraints" (Warner 1993, xi). Notably, he stresses that these concepts embed a heteronormative understanding of society. In a similar vein, I employ Warner's insights to explore the unrecognized constraints to our understandings of the normative force of our couple-familial oriented social models. Warner claims that queer politics must address the broader questions related to views of "social institutions and norms of the most basic sort" (ibid.). My line of thinking here builds on Lauren Berlant and Warner Warner's analysis of heteronormativity:

> A whole field of social relations become intelligible as heterosexuality, and this privatised sexual culture bestows on its sexual practices a tacit sense of rightness and normalcy. This

sense of rightness embedded in things and not just in sex is what we call heteronormativity. Heteronormativity is more than ideology, or prejudice, or phobia against gays and
lesbians; it is produced in almost every aspect of the forms and arrangements of social
life: nationality, the state and the law, commerce, medicine and education, as well as in
the conventions and affects of narrativity, romance and other protected spaces of culture.
(Berlant and Warner 1998, 548)

This important theoretical orientation presents us with the opportunity to think
about singlehood in broader social and political terms, and prompts the consideration
of issues related to social membership, identities, and normativity. Beyond this, it
creates a new agenda for singlehood studies, one which highlights singlehood as a
significant and unacknowledged aspect of social positioning. So, my argument is that
the study of singlehood provides us with novel and significant tools to explore not only
"What does it mean not to be in a couple?" (Budgeon 2008, 302), but—by extending
Warner's observations—creates the framework within which we can ask what singlehood can tell us about subjectivity and the categories of the social and the human.

Within this context, it is important to refer to Rachel Moran's (2004) argument that
the feminist movement has left singlehood off the feminist agenda. In a fascinating
historical analysis, Moran observes that "second wave feminism has failed to give full
recognition to single women as a distinct constituency with unique needs" (ibid.,
224–225). Feminism, she continues, has indeed lobbied for economic and political
equality and independence for women, yet seems never to have come to grips with the
possibilities of emotional individuality that are not incorporated within family and
marriage structures (ibid., 225).

I find Moran's observation applicable also to feminist theory and activism in general.
To a large extent, most feminist struggles take parental and conjugal ties as their points
of reference. Accordingly, in the final section of this book, I will revisit the political
and theoretical aims of this study. I do not merely call for acceptance or tolerance of
single women within a heteronormative, couple-oriented society but highlight new
modes of thinking, in which women can resist the narrow definitions of what is considered as women's appropriate conduct.

By adopting an interdisciplinary approach and integrating different theoretical
realms and perspectives, this book paves way for a new theorization of singlehood. To
accomplish this, a new conceptual groundwork is needed, within which singlehood is
moved from its simplistic temporary location and is understood in a wider social cultural context. That is to say, I conceptualize singlehood not merely as a troubling/
fascinating demographic phenomenon, a crisis, a social problem, or as anther sub-
category of family studies, but rather as a social phenomenon worthy of inquiry in its
own right.

Moreover, as a feminist sociologist, my interests are not confined to singlehood
alone. I think of singlehood as sociologically important, because it touches upon some
of the key questions in social thinking and raises pertinent questions about how people
make sense of their lives and organize their lives with others. A politicized analysis of
single living can open up and serve as a basis for advancing new visions of possible
subjectivities, communality, and sociability. A temporal reading of singlehood is an

important step in this direction, and the next section will develop the merits of this theoretical intersection further.

Theorizing singlehood and time

My first encounter with the sociology of time evolved, interestingly enough, from an attempt to collate different kinds of clichés ascribed to singlehood. Whilst doing so, I could not help but notice that one of the salient aspects of those clichés was time: "In the end she will die alone" was one, for instance; "What is she waiting for?" was another. People often comment that the single woman is about to "miss her train" or that she is "wasting her time." We ask single women if they are *still* single—and why; we also wish for them to get married *next* or *soon.* "Still; eventually; ever-after; waste of time; waiting; how long; when"—all these form part of the rich language of time.

As far as single women are concerned, time ever so often is perceived to be "on hold," "wasted," "empty," or "frozen." One can easily find cultural expressions that mock single women, characterizing them as overly selective, unable to make timely choices, and/or uptight and obsessed about getting married—an obsession which supposedly intensifies once they realize that "time is running out." When one compares the temporal notions of singlehood to those related to conventional discourses of couplehood, parenthood, and family life, a temporal hierarchy is revealed, one which distinguishes between those who are on time/off time, investing time/losing time, spending meaningful time/empty time, or controlling time/being controlled by time.

We often neglect to acknowledge that time is a socially constructed concept. However, social time is gradually becoming a significant conceptual category in critical thinking. Sociologists like Norbert Elias (1992) and Eviatar Zerubavel (1981) have studied how the invention of the clock and the calendar became a collective tool for time measurement imposing a secular time order.

As anthropologist Edmund Leach (1971) has noted, the regularity of time is not an intrinsic part of nature but rather a man-made notion which we project onto our environment for our own particular purposes. These devices endow society with different rhythms and measurements by dividing time into minutes, hours, days, weeks, and years. In this connection, Zerubavel notes that the socio-temporal order is "a socially constructed artifact which rests upon rather arbitrary social conventions" (Zerubavel 1981, xii). These symbols, according to time scholars, are significant tools for orientation, interaction, coordination, and regulation with which people establish orientation points along a continuum of change (Elias 1992; Zerubavel 1981).

As Émile Durkheim has pointed out, "a calendar expresses the rhythms of the collective activities, while at the same time its function is to assure their regularity" (Durkheim 2008, 10–11). For Elias (1992), time is not a personal reality but a collective one and, as he crucially asserts, although time feels private it is dictated by collective norms and forms. By the same token, Zerubavel has stated that "given its considerable temporal regularity, our social environment can easily function as the most reliable clock or calendar" (Zerubavel 1985, 14). Incorporating these important conceptual observations, I will challenge this seemingly private language of time, as well as its socially situated trajectories and identities.

Another significant point that should be made is that time is not singular, but multiple and heterogeneous (Adam 1990; Nowotny 1992). As Helga Nowotny notes, time "has many faces and assumes various shapes and forms of expression" (Nowotny 1992, 499). The temporal discourse of singlehood corroborates this assumption. It relies upon an abundance of metaphors, clichés, narratives, temporal concepts, and orientations, in which time moves quickly and slowly; is subjected to pauses and delays; or suddenly accelerates or takes unexpected turns, backwards and sideways. Often, single women have the experience of their time both running out and standing still at precisely the same time. Their time can simultaneously be perceived as empty, wasted, lost, and frozen. The different chapters in this book seek to problematize and contextualize these different temporal modes, and in that respect delve into the rich and multilayered temporalities which are reflected and produced by the category of singlehood.

Taking this into consideration, this discussion cannot limit itself to one, singular timescale which regularly evaluates women's social worth in accordance with normative prescriptions of linear trajectories embedded in heteronormative and reproduction regimes. As substantial feminist scholarship has shown, gendered perceptions of time are chiefly constructed by the biological deterministic arguments which perceive marriage and motherhood as women's primary life goals. Within this context, the discourse of the biological clock—so prominent in western societies today—reduces women's existence to features mainly articulated in biological and evolutionary terms.

Exemplifying this point, Merav Amir (2007) points out that the metaphor of the biological clock has become a new regulatory mechanism for producing gendered differences, and for disciplining single women to behave as timely feminine subjects. For Amir, the notion of the biological clock is embedded in essentialist assumptions of linear, goal-oriented, clock-driven temporal patterns, which individually and collectively impose fundamental constraints on women's lives. Evidently, this line of inquiry coheres with a long tradition of feminist criticism, which argues that women's subordination to men is sustained by beliefs of biological determinism. Clearly, one can easily detect the metaphor of the biological clock as having a hovering effect in single women's lives. However, my argument here is that a deconstructive study of singlehood and time must address the multilayered aspects of time, rather than a singular one. That is, stressing the role of the biological clock is merely a partial dimension of a complex phenomenon. Moreover, it is impossible to challenge the multi-layered and multi-formed temporal disciplining of single women by merely referring to the biological clock mythology as the only ideological order.

Following this line of thought, one of the objectives of this project is to problematize and broaden the scope of a new inquiry, which would subsume a wider range of temporal discourses, contexts, and concepts. Accordingly, my intent to deflate concepts such as *Wasting time* (Chapter 5) or *Waiting time* (Chapter 7), for example, does not accept them as given. Instead, it observes the social and cultural processes which produce them. Consequently, the chapters of this book are organized according to theses standpoints. Specifically, each chapter focuses on a different temporal concept, thereby acknowledging the rich and multifaceted temporalities which produce the category of singlehood and interpretations of time.

These issues will be addressed in more detail in subsequent chapters. For example, I will discuss how single women are perceived as failing to advance in a linear fashion and/or are accused of "wasting" time (Chapter 5), and are therefore designated with a waiting position within which their life is "on hold" (Chapter 6). An approach which acknowledges this diversity can provide us with a more nuanced understanding of how time-units, authoritative clocks, time-tables, and collective social rhythms formulate customary images of singlehood. In turn, temporal conceptions such as being late, being on time, time on hold, waiting and empty time all play an essential role in constructing these notions, and pave the way for a situated and relational reading of time.

Subscribing to the heteronormative temporal order

If we look at mainstream films, television series, advertisements, and global popular media in general, the figure of the single woman still represents a discursive unease (Taylor 2012), and serves as an easy target for social scrutiny, fear, and mockery. Indeed, the stereotypes of single women are mostly banal in their everyday presence. Female singlehood is still regarded as counter-normative, a deviant identity which will only lead single women to a disastrous future. These convictions are echoed in both Israeli and global media outlets. For example, in the winter of 2013, Elite-Strauss, one of Israel's biggest food manufacturers, launched a billboard campaign portraying an elderly woman with a chocolate bar in her hand. The slogan that accompanied this image stated: "Even if your granddaughter is still single, have a sweet day." This slogan, with its significant public presence via billboards across Israel's highway network, conveys a clear message: if one's granddaughter is still single, then there is a need for comforting and sweet consolation. From this perspective, singlehood can only be a temporary position; when it exceeds its temporal boundaries, it becomes cause for collective agony and distress. Herein, one's single status is not only a matter for private concern, but a collective one which positions both granddaughter and grandmother in a shared waiting position.

In a different clip produced as part of this advertising campaign, called "Sweetening it for Single Women," the same message is further promoted when Reut—a single woman—is offered a basket of chocolates to sweeten her single status. "Reut" garnered much public attention when a Facebook message—in which she wrote that she was looking for a husband—went viral. The clip features a conversation between Reut and the grandmother (the cartoon character from the billboard campaign) in which they discuss a list of possible marriage candidates for Reut. Towards the end of the clip, the grandmother narrates the story of an old woman who failed in love and is now left to die alone. The moral of the story could not be clearer.

This commercial campaign can be located within a global postfeminist climate, one in which irony and humor are used to advance conservative and traditional messages (McRobbie 2004; Taylor 2012). In this case, like many others, all the single woman can hope for is to unsingle herself (DePaulo 2006). Otherwise, she is warned, she will end up on her own just like that old woman who had no luck in her love life. These commercials are just two examples of the many textual artifacts which will be analyzed in this book.

Singlism in Israel is manifested through a rich repertoire of clichés, most of them with parallels in other languages. As an example, we can consider the manner in which single women are warned that they are about to "miss the train that everybody has already caught." "The train is departing soon!" or "The train won't wait for you and you'll be left waiting alone at the station!" they are told, again and again. I have heard these expressions in a variety of forms and versions. Many single women ask themselves, "Did I miss the train?," sensing that everyone around them is getting engaged and married. Images of old single women are often referred to as women "who have missed the train and now have no chance of catching the next one." To a certain extent, it could be argued that moving away from the linear, heteronormative expected life trajectory of marriage and parenthood are perceived as one's very own temporal miscalculations and failures.

The train is a key temporal metaphor in the discourse about single women, and not only in Israeli society. It represents a cosmic linear temporal order, upon which social order is established and regulated. In Zerubavel's (1981) words, this can be seen as our search for the temporal regularity which makes our life understandable. In this instance the train, the train tracks, and stations all symbolize the regularity of the temporal structure of our social life. Moreover, it also provides single women and their surroundings with the means of measuring their movements in time, and the extent to which they adhere—or not—to collective time schedules.

Thus, the fear of missing the train—like many of the examples discussed in this book—illustrates the ways in which collective schedules, clocks, and rhythms are translated and configured into an acute temporal awareness. Nonetheless, this temporal awareness is rarely problematized, and is left unquestioned in relation to representations of solo living. As Melucci (1996) stresses, our understandings of time are immediate and intuitive:

> Even when we understand immediately what we are talking about, we find it extremely hard to pin down what the experience of time actually means … in more ancient culture reference to time only conjured up a divine image—often a river god or another aquatic deity which, in the image of the flow, reflects the appearance and disappearance of things … the experience of time is characterized by a sense of thickness and a density that our definitions seldom provide and which, perhaps for this reason, cultures have sought to convey through the metaphor and myth. (ibid., 7)

In a similar vein, these temporal truths (such as "You will miss the train!") are rarely contested. In other words, they are formulated as objective facts reflecting the "real world" or the "facts of life," while neglecting the ways in which they are embedded in cultural practices and social relationships.

As my discussion in Chapters 5 and 7 will show, this rigid timetabling and scheduling is rooted in the sexist and ageist ideologies which imply that a single woman's market value declines with her age. These warnings are articulated as a wakeup call based on objective market calculations. Accordingly, above a certain age, single women have no chance but to adapt to this logic. The heteronormative message is also clear: if one fails to catch the train in time, there is no hope of getting married and fulfilling the injunction of reproductive continuity. The consequences of such belated rhythms are social marginality and exclusion. This is one of the many examples which demonstrate the links between the social organization of time and relations of power and social control. Undoubtedly, conceptions of successful timing and time management are based on compliance for one's continued existence, otherwise one runs the risk of becoming an "old maid," a "crazy cat lady." Thus, missing the train infers that there is no chance of becoming respected female subjects, achieving full membership in a society articulated in familial, heteronormative terms. The extent to which these temporal truths and hierarchies are internalized by single women and their surroundings cannot be undermined.

It should also be noted that the theoretical lens offered in this study enables us to be attuned to what single women in Israel say about time, as well as to the temporal identities ascribed to them in this discursive process. In that respect, I have found Ramón Torre's (2007) theoretical and methodological advice particularly inspiring. Torre notes: "The clarification of time must always take into account what the social agents say or assume about time: their lexicon, their 'grammar,' their images, and even their ambivalences and inconsistencies" (158–159). Endorsing Torre's advice, and being attuned to the texts analyzed in this book, opens new directions of thinking about time and singlehood beyond clock time or linear heteronormativity.

However, the metaphors used may hide or highlight certain aspects of the singlehood experience, and their conventional perceptions. As Lakoff and Johnson point out:

> The very systematicity that allows us to comprehend one aspect of a concept in terms of another (e.g., comprehending an aspect of arguing in terms of battle) will necessarily hide other aspects of the concept. In allowing us to focus on one aspect of a concept (e.g., the battling aspects of arguing), a metaphorical concept can keep us from focusing on other aspects of the concept that are inconsistent with that metaphor. For example, in the midst of a heated argument, when we are intent on attacking our opponent's position and defending our own, we may lose sight of the cooperative aspects of arguing. Someone who is arguing with you can be viewed as giving you his time, a valuable commodity, in an effort at mutual understanding. But when we are preoccupied with the battle aspects, we often lose sight of the cooperative aspects. (Lakoff and Johnson 1980, 458)

By exploring the hidden meanings of temporal metaphors such as the ticking of the biological clock or "missing the train," one can think of alternative temporalities. In other words, "missing the train" might actually present the opportunity for something else. By subverting these metaphors and turning them on their heads, one has the opportunity to reflect more closely on non-linear trajectories, and the liberties that are granted by having time beyond or outside the normative track. Counter-hegemonic timetables, such as the benefits of waiting and even in some cases preferring to miss the train can enable women to move beyond what Halberstam terms as "conventional logics of development, maturity, adulthood and responsibility" (Halberstam 2005, 13). Such alternatives to the temporal heteronormative framework will now be discussed.

Counter-representations of long-term singlehood

The heteronormative scripts about female singlehood are so deeply embedded in our social imaginary that it seems almost impossible to contest them. However, numerous internet sites, personal blogs, and local initiatives have sought to debunk common understandings and stereotypical attitudes towards single men and single women. In what follows, I will examine some of the alternative voices offering subversive views of female singlehood and gendered temporal timetables.

In a *New York Times* cover story published in February 2015, Emma Morano, the oldest woman in Europe and the fifth oldest person in the world, noted that one contributory factor to her longevity was being single. Morano's story was published under the suitably catchy headline, "Raw Eggs and No Husband Since '38 Keep Her Young at 115" (Povoledo 2015).

"I didn't want to be dominated by anyone" (Davies 2015), Morano explained, thus crediting her longevity to the fact the she did not re-marry after separating from her husband in 1938. The *New York Times* piece (Povoledo 2015) went viral, receiving extensive media coverage. But Morano is not alone; 109-year-old Jessie Gallan from Scotland, for example, revealed when interviewed by the *Daily Mail* (2015) that her "secret to a long life has been staying away from men. They're just more trouble than they're worth," she added, saying that "I also made sure that I got plenty of exercise, eat a nice warm bowl of porridge every morning and have never gotten married."

Various bloggers soon recognized the potential that these stories possessed for challenging some of the well-established discourses of singlehood. For example, Chrissa Hardy (2015), a blogger writing for Bustle argues that we should pay attention to Morano's views on romance: "The fact that she was able to put her needs first and end a relationship in which she was no longer happy says a lot about the kind of boss lady Morano has always been." According to Hardy, Morano "values her freedom, and she is perfectly comfortable with the life she has built since" (ibid.).

Thus, Morano's story leads Hardy to reach the following conclusion:

> So instead of wallowing about your lack of Valentine's Day (or Singles Awareness Day) plans yesterday and whether or not you'll end up finding "The One," think of Emma Morano, and how her long and happy life has been centered around her romantic freedom. She is living proof that a husband is not the key to eternal happiness for everyone, and that you should find what works for you and stick with it. (ibid.)

I concur with Hardy's reading. Morano and Gallan's stories, similar to many alternative scripts advanced by single women, carve out their own time path and life-course trajectory. Such paths are still rarely recognised in mainstream society, and reflect the need to harness conventional hegemonic discourses. Indeed, Morano and Gallan do not define themselves through "the love plot of intimacy and familialism that signifies belonging to society in a deep and normal way" (Berlant and Warner 1998, 554). Neither do they adhere, as Hardy points out, to the conventions of the "happily ever after" script; in this way, they show the possibilities that exist for resisting the regulatory effects of heteronormative time.

Such accounts also point to the possibility that singlehood is both a social category and an analytical tool for questioning some of our core understandings of the normative and the natural. Morano and Gallen's biographies, as presented above, echo some of the accounts presented in this book. These accounts do not follow the heteronormative linear trajectory, in which marriage and procreation are considered as obligatory milestones (see Chapter 2); neither do they define their lives as frantically waiting and searching for "Mr. Right" (see Chapter 7). On the contrary, Morano and Gallan tell us that their single life trajectory has provided them with longevity, health, autonomy, and freedom.

In Chapter 8, I contend that one of the common temporal scripts that single women are expected to identify with is of being miserable on Valentine's Day, or as Hardy terms it, "Singles Awareness Day." However, Hardy also provides an alternative script, in which one can appreciate one's romantic freedom and envision a different kind of futurity. Murano's story enables Hardy to envision a different life narrative, which neither follows the "happily ever after" life script, nor embraces the script that ends in catastrophe. As she emphasizes, Murano's biography reflects that "a husband is not necessarily the key to eternal happiness." Her suggestion could be read as perhaps providing a different set of "happy endings" scripts. This alternative storyline echoes Sara Ahmed's (2010) work, which explores the ways in which expectations of happiness operate as a regulatory brake, one which prioritizes normative ways of lives while precluding others. In this connection, Ahmed writes, one of the primary social indicators of happiness is marriage, which comes to represent "the best of all possible worlds" (ibid., 6). She further notes that this serves as an example of the ways in which happiness is used to reconfigure social norms as social goods, as well as restricting ways of imagining one's future.

Building on Ahmed's analysis, it could be argued that happiness is out of reach for the "miserable old maid," in common with representations of the queer, the migrant, or the feminist killjoy discussed in Ahmed's book. According to this line of analysis, the ever-single woman is excluded from the joys of conjugal and familial life which come to represent "that to which good feelings are directed towards" (ibid., 21). Moreover, the ever single woman is identified with the impossibility of happiness, an unhappiness which is perceived as a failure and as a deviance.

Thus, Hardy opposes the very exclusion of single women from the possibility of happiness and the way futurity is normatively envisioned. Moreover, if we follow Ahmed's line of analysis, we can argue that Murano's storyline questions the very process through which the norm of marriage becomes a social good and indicator of the good life. This line of analysis could thus also illuminate why long-term singlehood has political potential to challenge the common-sense norms regarding what makes life meaningful and valuable, and the norms that are ascribed to happiness and normativity. Singlehood is understood as an alternative way of living, which is chosen from a profound sense of awareness and self-determination.

In this context, this project aims to challenge some of the binary models that dominate current thinking about singlehood and family life, differentiating between happiness and misery, loneliness and togetherness, health and pathology. By the same token, I have cautioned elsewhere (Lahad 2014) against the simplistic embrace of the "chosen singlehood" formula and its possible implications for novel forms of the politics of identity and recognition. I have argued that we should take into consideration the multiple experiences of women's lives, which should not be reduced to the choice/non-choice, happy/unhappy single woman dichotomy. I contend that the single-by-choice formula can obscure and delegitimize dualities, contradictions, and complexities.

This line of analysis builds on Reynolds's (2008) work, which provides a nuanced reading of single women's lives. In her studies, Reynolds has examined how single women juggle their repertoire of choices and chances and consequently view choice not as a factual notion, but rather as one of the discursive resources available to single women. Significantly, the choice discourse endorses the deeply rooted binary thinking which precludes other potential discourses on inconsistency, hesitation, ambivalence, and confusion.

I have also argued that the new images of liberated, empowered, freely-choosing single women could essentialize women's lives, and consequently constitute new hierarchies between those who can and those who cannot follow the dictates of the new regime of chosen/non chosen singlehood or the miserable/happy single. These types of classificatory categories create fixed and static boundaries, with limited possibilities for slippage between these poles.

Women's identities are connected to class, age, religion, ableness, and sexual orientation; all of these factors enable and narrow one's options for holding on to the position of chosen singlehood. Moreover, an intersectional perception of singlehood should take into consideration the fact that the identities of single women are connected to different experiences and changing discursive and material conditions. For example, singlehood is classed, and the single-by-choice discourse cannot be based solely on white middle-class female experiences to the exclusion of other women. While middle- and upper-class single women may have the material conditions for choosing singlehood and enjoying time on their own; such access for living alone or having time alone might not be as readily accessible for single working-class women. Viewing singlehood solely through the theoretical framework of individual choice, self-determination, and personal happiness could lead us into a conceptual dead-end, preventing us from developing a more political and nuanced understanding of singlehood. Bearing this in mind, my criticism does not aim to discount the importance of articulating singlehood from a confident and unapologetic position, as well as establishing new models which enable long-term singlehood as a viable and desirable life category.

However, despite the counter-representations of singlehood presented above, the negative stereotypes about single women are strong and remain common in many societies, particularly, as I will show, in Israeli society. This then leads us to ask why this stigma is still so widespread and what bestows it with so much force? This question will be addressed throughout the book. Before winding up, the concluding section of this introduction lays out some of the methodological considerations and the social context for of this project.

The context for this study

Israel presents a fascinating case study that can help deepen our understanding of singlehood and temporality, particularly due to what has been termed as the traditionalism–modernism paradox of contemporary Israeli society (Bystrov 2012). On the one hand, Israel has undergone dramatic transitions in family life. In common with many European and American societies, the country has been affected by societal trends such as: the multiplicity of living arrangements; postponement of the age of marriage; rising rates of divorce (for example, in England in 2015 the divorce rate per 1,000 married men and women was 9.8, while in Israel in 2013 it was 9.1; see Bingham and Kirk 2015; Central Bureau of Statistics 2015); LGBT partnerships; single-parent families; and single-person households. On the other hand, despite these far-reaching changes, familism and traditional gendered expectations towards women prevail (Fogiel-Bijaoui 1999).

The centrality of family ideology and relatively high birthrates in Israel are perceived as related to various factors, such as: the "demographic war" to keep the Jewish popula-tion as a majority group; the effects of the Holocaust; the role of the religious establish-ment in the political and cultural system; and religious Jewish practices and beliefs aimed at enhancing the Jewish character of the state of Israel (Portuguese 1998). It should be mentioned that the fundamental place of Jewish religion in Israel is the reason why personal status is regulated through religious law. The obligation to be a mother is also present in religious commandments, such as "be fruitful and multiply," which have been given secular ideological validity as well (Donath 2015).

The formation of a large family is still considered, in many ways, to be a patriotic act and part of the national mission (Fogiel-Bijaoui 2002). The family-centered order of Israeli society is manifested, for example, in welfare policies, family allowances, and generous state funding for infertility treatment technologies (Portuguese 1998). For Portuguese, the signs of Israeli familism are easily detectable: Israeli women marry relatively earlier, bear more children, and divorce less than their counterparts (ibid.). The centrality of family in Israeli society today is also reflected in findings emerging from an impressive body of scholarly writings that have examined Israel's pro-natalist ideology and policy (Berkovitch 1997; Donath 2011; Hashiloni-Dolev 2007; Shalev and Gooldin 2006). In another study, Don Handelman also pointed out that the meta-phor of the family in Israeli society is central to the construction of the national imagi-nation. The nation, like the family, is perceived as an organic entity, Handelman writes, and Israelis correspondingly are imagined as one big family (Handelman 2004, 13).

Relatedly, *motherhood* is considered the most precious quality in women's lives, a significant indicator of women's inclusion in Israeli society, as well as an important avenue for collective belonging (Teman 2010). Thus, to a large extent women are constructed first and foremost as wives and mothers (Berkovitch 1997), whose most important obligations consist of bearing and rearing children. Through this formulation, marriage and motherhood are rendered as intelligible forms of subjectivity, in turn construing dominant sets of hierarchies and the normative codes of an "imagined normality."

In this social setting—which demands that a woman be a wife and mother—single and childless women are the object of a constantly scrutinizing gaze, which creates a self-policing subject. Thus, and in the face of fundamental changes to the availability of reproductive choices, women who have chosen not to have children are still subjected to hostility and social disapproval, and are heavily stigmatized (Donath 2011). According to Orna Donath's (2011) study, Israeli women who do not take part in this venture are, to a large extent, still considered to be abnormal and therefore cannot "really" choose this life path.

Within this context, the category of chosen or long-term singlehood of women is rarely presented and legitimized. This is one reason why most of the Israeli texts analyzed in this study echo these relatively traditional views about motherhood and marriage. However, my study also reflects some of the new oppositional voices attempting to challenge the limiting stereotypical representations of single women and the hegemonic ideals of family life. These voices indeed cohere with women and gay liberation movements, who have led changes in cultural attitudes and expectations of Israeli women today. Indeed, as some studies have shown, these changes have led to a growing acceptance of divorce and single motherhood. This is not to say to say that the stigmas directed towards single mothers have vanished completely, but there are indications of certain shifts in the attitudes of the secular Jewish population towards single motherhood (Hashiloni-Dolev and Shkedi, 2007; Lahad and Shoshana 2015; Landau 1996). What distinguishes Israeli society from other societies is the significance of reproduction in Jewish culture (Kahn 2000; Lahad and Shoshana 2015); this cultural climate poses significant discursive barriers to the acceptance of childless single women in Israel.

Methodology

Drawing on a wide variety of Israeli cultural resources I will attempt to sketch some of the meaning-making processes of singlehood and time. In this context, it is worth mentioning that the various texts under examination are viewed as cultural sites, in which the discursive construction of the socio-temporal aspects of singlehood are reflected and produced. That is, the selection of data for this study stems from the contention that popular culture, everyday talk, and new media technologies affect, sustain, and alter the deeply ingrained understandings through which singlehood is constituted and formed nowadays.[8] The methodology and choice of materials is closely linked to these rapidly changing social realities. In other words, this study is attuned both to local-global discursive formations, and to the old-new contexts which constitute and represent contemporary understandings of singlehood and social time.[9]

This study employs a qualitative content analysis-based approach to explore the relevant themes that link the discursive categories of singlehood and time. My choice of Israeli internet columns written by and about single women, clichés, commercials, and popular articles is related to my contention that these sites convey deeply ingrained socio-temporal norms, with which the cultural tag of singlehood and representations of single women can be further interpreted. The majority of the texts are columns published on the Israeli portal *Ynet*, as I consider it one of the principal websites on which themes of singlehood, dating, and relationships are discussed.

However, during the years that I have researched singlehood, the online sphere in Israel has expanded significantly, with web platforms such as *Saloona* and *Tapuz,* as well as news websites like *nrg, Mako, Haaretz* and *The Marker*, publishing columns and articles which touch upon these issues. Indeed, this corresponds with the global tendency, within which a wide variety of internet portals, blogs, and forums have shown growing interest in single women's lives and singlehood in general (Taylor 2012).

Ynet currently remains the central platform within which issues related to singlehood, dating, and relationships are discussed, almost on a daily basis. This is why I visited *Ynet* (www.ynet.co.il) daily between 2006 and 2016, looking for columns discussing the lives of single women.[10] The texts examined were either personal columns written by single women recounting different aspects of their singlehood, or texts written by dating and/or relationship advisers who contribute regularly to *Ynet*. Most of the texts selected for analysis were chosen from a subsection in *Ynet* entitled "Relationships," where references to discussions about late singlehood (amongst other things) are discussed. This subsection contains various topics like "Dating," "Weddings," "Couples," "Pride" (LGTB discussions), and "Sexuality," alongside personal web columns and dating and relationships advice.

I have also visited other web platforms on a weekly basis, by using the Israeli google search engine and snowballing blogs and social media debates by typing phrases such as "single woman," "single women," "old/aging single woman," "extended singlehood," "late singlehood," and "long-term singlehood." I have taken into consideration several variables concerning the genre of the text (a blog, a column, a news or life-style column) and the author's position (a journalist, blogger, or a person interviewed for an article).

The columns analyzed form part of a flourishing Israeli internet culture, in which questions regarding personal relationships, dating, and single life come to the fore. I approach the web columns as a rich source of data, particularly as this medium has become a popular outlet of expression for both readers and writers. In this respect, I view the columns written by and about single women as forming an important global-cultural space for interpretation and debate. As will be demonstrated throughout this book, discourses on single women and time cross national boundaries, and in this sense demonstrate the shared temporal understandings of singlehood across the globe.

Hence, I occasionally analyze texts outside Israel for a couple of reasons. First, Israeli culture publishes translated texts and screens many international box office hits and popular television series revolving around singlehood and single women (e.g., *Sex and the City*; *Girls*; *The Bachelorette*; the *Bridget Jones* films). In effect, as media scholars have shown (Taylor 2012), this globally mass-mediated imagery has changed the creation and circulation of discursive constructions of singlehood and this certainly applies to the Israeli context. Various studies have explored the Americanization of Israeli society and the ways in which many Israelis are fascinated with the "American way of life" and American culture. (See, for example: Aronoff 2000; Avraham and First 2003; Ram 2013)

The Israeli media regularly refers to and translates columns and articles relating to female singlehood that have attracted media attention on a global scale, most of these articles being from the US and the UK.[11] The global proliferation of the discourses of singlehood is especially pervasive, particularly taking into consideration just how open the Israeli media is to global influence. In this vein, *Ynet* is representative of the way in which cultural artefacts operate on a global scale.

Second, I attempt to show the similarities between the Israeli discourse and discourses outside the Israeli context. Third, most texts still echo a traditional attitude towards singlehood, reflecting hegemonic gendered perceptions. One of my objectives is to show alternatives to these temporal configurations and the ways in which female subjects are portrayed. As singlehood in the Israeli context has almost escaped politicization—for the moment—most examples of such discursive alternatives are to be found in studies and texts published outside Israel.

Drawing on Foucauldian discourse analysis methods (Foucault 1972) and feminist discourse analysis (Lazar 2007), my objective is not just to understand the mechanisms that construct the cultural tag of single women, but also to deconstruct some of the underlying premises and regimes of truth. In other words, my reading of the texts derives from an attempt to locate these cultural schemas in a specific historical moment and understand their wider social and gendered contexts. As Michelle Lazar (2007) elaborates: "the aim of feminist critical discourse studies is to show up the complex, subtle, and sometimes not so subtle, ways in which frequently taken-for-granted gendered assumptions and hegemonic power relations are discursively produced, sustained, negotiated, and challenged in different contexts and communities" (ibid., 142).

Discourse, according to Michel Foucault, "should constitute thought, clad in its signs and rendered visible by words or, conversely … the structures of language themselves should be brought into play, producing a certain effect of meaning" (Foucault 1972, 227). Furthermore, he emphasizes, discourse should be perceived "as a violence that we do to things, or, at all events, as a practice we impose upon them; it is in this practice that the events of discourse find the principle of their regularity" (ibid., 229). The underlying premise of critical discourse analysis is that discourse shapes reality in accordance with the ideological interests of social groups. More specifically, this approach seeks to determine "what structures, strategies or other properties of text, talk, verbal interaction or communicative events play a role in these modes of reproduction" (Van Dijk 1993, 250).

In other words, the aim of critical discourse analysis is to shift the focus away from the "objective" or "essential" qualities of a text, and towards a reading that reveals the randomness, arbitrariness, and social construction of reality. I therefore consider the web columns to be compelling sites for understanding how taken-for-granted cultural constructs of singlehood are represented and produced. Moreover, the different columns are also considered to be the site of a cultural struggle (Fiske 1996; Lazar 2007), which also offers a unique prism with which to understand how current cultural meanings of singlehood, time and feminine subjectivity are validated or contested.

Outline of the book
Chapter 2: The linear life-course imperative

Chapter 2 opens with what I consider to be two important temporal conceptions in the social interpretation of female singlehood: the belief in progressive linearity and the heteronormative paradigm of the life course. The approach guiding my analysis integrates a social constructionist lens as well as recent theoretical developments in feminist and queer time studies, which challenge the heteronormative life course. Building on these perspectives, this discussion demonstrates the ways in which linearity and its related concepts such as progress, reproduction, and continuity are socially and ideologically situated. A critical discussion of the linear temporal order serves as a point of departure for this chapter, which will be followed by examining its normative implications for female singlehood.

In the second part of the chapter, I analyze the life course scheme, which I consider to be a major conceptual paradigm through which late singlehood is judged and evaluated. Thus, I make the case that the essentialist and naturalized life-course paradigm is a particularly powerful cultural template, but one that is rarely criticized in popular and scholarly discourses on singlehood and is taken as a given. Thus, instead of adhering to the prevailing normative linear paradigm of the progressive life-course order, I ask to critically re-evaluate its terms, convictions, and powers.

Chapter 3: Singlehood as unscheduled status passage

Chapter 3 expands the analysis of the expected linear life-course trajectory from a different perspective. The focus of this chapter is a conceptual analysis of *becoming single*. I also explore the discursive mechanisms that constitute it as a biographical disruption. I argue that this process is rarely problematized in relation to singlehood or figured as a default life trajectory. My discussion examines this path in relational terms, in which the process of becoming single and the transition from normative to late singlehood is produced by socio-temporal truth statements. Thus, the stages of singlehood—or more specifically what I term as the *singlehood career*, drawing on Goffman's use of the term—comes to existence through a hegemonic temporal gaze. Throughout this chapter, I show how this gaze is established through social interaction and is ingrained in collective social perceptions. In this chapter, I also demonstrate how becoming single is a subtle non-institutionalized transition process, in which the entry and exit from "normative singlehood" to "late singlehood" occurs without rituals or official formalities.

The second part of this chapter offers a temporal analysis of the question "Why is she *still* single?" as signifying the transition to late singlehood. My intent here is to explain the discursive formations and implications of this ubiquitous question, and to shed light on how popular knowledge about single women is produced and circulated. Thus, I do not ask why single women are single, but rather examine how the question itself is discursively constructed in relation to how singlehood is figured as an unscheduled trajectory.

Chapter 4: Facing the horror: becoming an "old maid"

My analysis of time continues by exploring the temporal category of age. Incorporating recent literature on age, feminist theory, and singlehood, this chapter re-evaluates the image of the "old maid" alongside the omnipresence of age, sexism, and ageism in current discourses on female singlehood in Israel. It asks, what gives this powerful stereotypical image so much discursive force and makes it so defiant to resistance and deconstruction? I find that questions such as "Why are twenty and thirty-plus single women depicted as old?" and "Why are thirty-plus married mothers represented as 'young mothers?'" are illustrative questions which emphasize that single women are aged by a culture determined by socially framed expectations. Drawing on this, I also wish to understand the discursive process that causes single women to "age faster"; *how* do single women "age" differently from coupled and married ones? These queries reflect our line of inquiry, which views aging as a gendered and a heteronormative based process.

These questions are discussed by exploring how the predominant cultural perceptions of age appropriateness, age segregation, age norms, and ageism play a crucial role in the construction of late singlehood and gendered timetables in general. My contention is that single women are faced with a triple disfranchisement, based on age, gender, and single status. Given this, I argue that single women undergo a process of *accelerated aging*, leading towards their *social death*. Thus, this chapter also makes a significant contribution to critical age studies and feminist age studies, by reworking these categories and opening up new ways to critically revisit the authority of age and sexist and ageist practices. It also points out that ageism and age-based discrimination do not necessarily apply merely to the social category of old age, but are practiced at different phases of the life course.

Chapter 5: On commodification: from wasted time to damaged goods

While the previous chapter focused on how ageism, sexism, and singlism coalesce, this chapter covers an in-depth analysis of how they come together and are discursively articulated through the commodified language of time. Chapter 5 includes various examples in which singlehood and single women's subjectivity are constantly measured through the conditions of the marketplace. Single women are perceived as having a short shelf-life, and above a certain age will be considered as damaged goods and consequently warned "not to waste their time," and "to find a husband before it's too late." The reoccurrence of these concepts have prompted me to examine how this temporal economic language provides a set of powerful presuppositions, through which single women are objectified and ascribed with an inferior social status. My discussion here considers this to be a significant discursive context, through which single women's oppression occurs and hierarchical gendered power relations are sustained.

This line of inquiry allows a rich analysis of how this age- and gender-based temporal logic conjoins with the rhetoric of supply and demand. In this chapter, my aim is to create a new understanding of what are considered as undisputed market laws, and the way time is reified and commodified. Following this line of thought, this discussion raises questions such as: To what extent does the abstraction of time act as a quantifiable measure that controls the lives of single women? In which ways does the commodified language of time set up our perceptions of female singlehood? What are the discursive mechanisms through which single women are considered as damaged goods, stamped with expiration dates? And lastly, how are temporal practices, such as *wasting time* and *accumulating time*, reconfigured in relation to single women's time? The discussion of these questions aims to set a broader perspective and provide alternative ways of thinking about singlehood. In particular, it seeks to disconnect this temporal discourse from normalized concepts of market logic, exchange value, consumer goods, and the assessment of women's ability to be "sold" and "traded."

Chapter 6: Taking a break

Chapter 6 gives us the opportunity to discuss temporal rhythms of daily life from a different perspective. In the first part of the chapter, I critically assess normative social rhythms by discussing temporal concepts such as *timeout* and *taking a break*, as well the implication of *breaking away* from the linear trajectory. The textual analysis of web columns written by and about single women reveals that taking a limitless break from the non-stop search for a husband is considered as a disrupting act, which might prevent them from attaining a reproductive and meaningful futurity.

According to this view, engaging in a non-stop search for a husband is regarded as a productive and required temporal trajectory. Drawing on Hogne Ÿian's (2004) sociological study of time, I argue that what is considered as a temporal *timeout* can turn into a permanent *dropout* from the collective linear trajectory, with limited chances of rejoining it. I suggest that the demand for a timeout is also an act of resistance, an attempt at breaking away and taking a timeout of time and therefore a subversive practice which conveys a claim for temporal agency and control of one's time.

In the second section of this chapter, I take this line of analysis further by exploring images of *time on hold* and *frozen time*. Central to this analysis are also questions of mobility, speed, and temporal subjectivity, which lead me to examine how and why single women are figured as immobile subjects frozen in time. According to these texts, single women are viewed as trapped in their own immobility, a temporal position in which they have lost their telos and agency. This leads to a discussion of what is figured as a breakdown in the articulation of time (Reith 1999) and arrested flow of time which disconnects the present from the future and empties them out of meaning and substance.

Chapter 7: Waiting and queuing

This chapter is devoted to a critical analysis of the temporal construct of "waiting." I analyze representations of waiting in everyday clichés, commercials, popular songs, and web columns, as well as representations of bridesmaids in popular culture. The figure of the single woman waiting to enter coupledom and married life has become deeply embedded in conventional thinking about single women, and these representations offer a useful case study as they highlight the temporal organization of social life and the related relations of power structures. From this viewpoint, waiting is examined as an interactive setting, representing and producing rigid societal timetables, as well as traditional feminine subjectivities.

Drawing on Leon Mann's (1969) and Barry Schwartz's (1975) observations of queue culture, I propose to examine representations of single women also as queuers standing in line, waiting to enter matrimony. Observing and interpreting this social interaction as a queue offers multiple dimensions of analysis. For example, a temporal analysis of the queue as a social microsystem lends insight to how temporal norms and temporal mechanisms are established. From this perspective, the status of single women can be measured according to their location within and outside what I term as a *heteronormative queue*. In much the same way—and by extending Pierre Bourdieu's (2000) analysis of waiting—I examine how waiting is both an exercise and effect of power. I argue that these sets of images constitute compliant temporal subjectivities, which integrate them into an unquestioned heteronormative order.

Chapter 8: Time work: keeping up appearances

Continuing and extending my focus of the interactional aspects of single temporality, Chapter 8 develops these aspects further. In the texts analyzed in this chapter, single women reveal their hesitations about, and the obstacles of, being in public on their own. This chapter offers a temporal reading of everyday social interaction by employing a Goffmanian analysis of the ways in which single women prepare themselves for social interaction by taking into consideration temporal dimensions in an attempt to control their impression management. Concepts such as *participation units*, *loss of face*, *civil inattention*, and *impression management* are used to examine the temporal dimensions of the presentation of the single self in public. In this way, I draw attention to the temporal context within which social interaction takes place and demonstrate how time marking assumes overwhelming importance.

In this chapter, I also rely on Durkheim's (2008) and Zerubavel's (1981, 1985) works on temporal demarcations. By examining the social meanings attached to time units such as night and day, the week, and the weekend, every day and holidays, I argue that these time conventions have an important bearing on the single woman's opportunities for appearing and interacting in public. It is argued that the temporal interpretations of time during holidays like New Year's Eve, Valentine's Day, and social occasions like dinner time have an important bearing on the single woman's visibility, and impact her ability to orient her appearance, and consequently her sense of self agency, in public settings.

Chapter 9: Discussion: another time

I conclude the book by analyzing possible alternatives to the hegemonic temporalities explored in the previous chapters. I argue that these accounts present us with alternative perceptions of temporality, through which women can articulate their own ideas about time and their single status. In this chapter, I am particularly interested in exploring how resistance to time norms is represented, and how women attempt to *reclaim their time* and destabilize common-sense life paths and schedules. In this way, their claim subsumes a sense of controlling time and a way of attaining *temporal autonomy and agency*. By exploring resistance by single women in Israel and elsewhere in the world, I seek to explore how a new agenda for singlehood studies can be formulated, and singlehood itself understood in broader political terms.

Notes

1   According to the Israeli Central Bureau of Statistics (2015) which defines women as over 15 years old and according to the Office for National Statistics (2015) of the UK, which defines women as over 16 years old.

2   For an excellent review of Intersectional Feminist theory, see Oleksy (2011).

3   For an interesting discussion of religious singles in Israel, see Engelberg (2011, 2013).

4   See, for example, Byrne and Carr (2005); Budgeon (2008); Cobb (2012); Dales (2013, 2014); DePaulo (2006); Klinenberg (2012); Jamieson and Simpson (2013); MacVarish (2006); Nakano (2011, 2014); Reynolds (2008); Simpson (2006); Song (2014); Taylor (2012); Trimberger (2005); Wilkinson (2014).

5   See, for example, DePaulo (2006); Trimberger (2005).

6   To the best of my knowledge the only work that has pointed out the importance of a temporal reading of feminine singlehood is Negra (2009) in a short discussion of representations of single women in popular culture.

7   See, for example, Cobb (2012); Evertsson and Nyman (2013); Jamieson and Simpson (2013); Klinenberg (2012); MacVarish (2006); Nakano (2011); Reynolds (2008); Song (2014); Taylor (2011); Wilkinson (2014).

8   These observations also rely upon a rich and varied scholarship on popular culture, which views it as not only a key site for formation of identities and everyday realties, but also an arena in which consent and resistance are intertwined. For more, see Hall (2010); Illouz (2007).

9   While most sociological work on singlehood today focuses on in-depth interviews, the present study seeks to add to the existing literature on singlehood and social time, by incorporating these new global shifts and translating them to new research questions. For studies based on in-depth interviews, see, for example, Budgeon (2008); Byrne (2000); MacVarish (2006); Nakano (2011); Reynolds (2008); Simpson (2003, 2006); Trimberger (2005).

10   According to a recent survey (Goldenberg 2015), *Ynet* is the sixth most visited website in Israel (following the English and Hebrew versions of *Google, Youtube, Facebook,* and *Walla*).

11   For example Lorri Gottlieb's (2008) article published in the *Atlantic* was discussed in *Ynet* (Regev 2010). Or more recently Kate Bolick's (2015) book *Spinster: Making a Life of One's Own* was analyzed in *The Marker* (Shechter 2015). Her article published in the Atlantic (Bolick 2011) was also mentioned in the Israeli media (Sa'ar 2012). Likewise, Rachel Greenwald's (2004) book *Finding a Husband After Thirty-Five: Using what I Learned at Harvard Business School* was translated into Hebrew and widely discussed in the Israeli media. For more see Lahad (2007).

2

The linear life-course imperative

One of the more prevalent clichés in Israeli culture is the consolation, "By your wedding day you will feel better." This sentiment is often directed towards small children and is intended to be both comforting and hopeful at the same time. The sentiment not only assures children that with time they'll feel better; it also constantly reminds them of their prospects for the future. In fact, it leaves no room for doubt regarding the heteronormative life-course trajectory, one that leads—eventually, but inevitably—to marriage and children.

Given that marriage is an important milestone, a turning point in one's life course, it is no surprise that similar versions of the same consolation can be found in German, Russian, Spanish, and Portuguese cultures, for example. Indeed, many young girls dress up during Halloween and Purim (a Jewish religious holiday) as brides; the popularity of the "dress up as a bride" game endures. In a similar vein, small girls around the world play with Barbie bridal outfits, and many of the toys marketed at girls, including domestic appliances and little baby strollers, predict a heteronormative, reproductive future. In romantic comedies and television commercials, one often finds sequences where single women reflect on how as young girls, they had already planned their perfect wedding.

Reflecting further upon the consolation: beyond the belief that time heals everything, the wedding is positioned as an indisputable milestone, a life goal, an important transition point in one's life course. Additionally, popular clichés like these guide the child's future life trajectory by emphasizing a linear, heteronormative progression, along which the ultimate destination is visible, clear, and certain. This conceptual temporal frame guides the upset child towards an imagined and desirable future. A prominent cultural channeling process is revealed here, one which predetermines the child's future performance and reflects a collective life scheme. The reassuring and comforting promise predicts an identifiable heteronormative future, one embedded in a set of dominant expectations regarding the valuable and desirable in one's future life course.

In this chapter, I pay close attention to the ways in which this linear *telos* constitutes some of the major discursive frameworks of the single woman's life trajectory.

Life-course research has developed extensively since the late 1960s, most notably in the fields of developmental psychology, but also in contemporary anthropological, sociological, and gerontological studies. The resultant abundance of scholarly literature offers a wide variety of conceptual models that simultaneously reflect and produce some of our most pervasive cultural understandings.

My approach to the study of the life course in this chapter integrates a discursive and social constructionist lens, whilst also paying attention to recent theoretical developments in queer time studies. Consequently I view the essentialist and naturalized life-course paradigm as a particularly powerful cultural template. More specifically, my line of research draws upon scholarship which views the notion of the life course as a potent, socially constructed metaphor (Becker 1994; Holstein and Gubrium 2000). James Holstein and Jaber Gubrium, in *Constructing the Life Course*, point out that the life course is a representational tool, crafted and used in the process of interpreting personal experiences through time (Holstein and Gubrium 2000, x).

This line of analysis attempts to view the life-course construct as an interactional accomplishment; a social form according to which individuals make sense of their everyday lives. This perspective is also reflected in Martin Kohli and John Meyer's (1986) work, which views the life course as a persuasive cultural institution and as an age-graded structure producing time-ordered opportunities and constraints. My analysis views and joins this literature which examines the life course as an ideological—and therefore, fabricated—discourse of existence, including the developmental sequences which are commonly treated as objective and natural features of life.

Such considerations also highlight the manner in which the discursive formation of the life course is standardized and made uniform by what Kohli has described as a *life-course regime* (Kohli 2007). This regime defines the life course as comprised of rigidly defined sequential developmental stages with administrative rulings (ibid.). According to this perspective, conventional life-course patterns are in-and-of-themselves social constructs, the consequence of a dominant discursive practice that functions as a core regulator of formal and informal social laws. Life-course models, as Gay Becker notably explains in her study on disrupted lives, serve as the basis for the development of cultural models of how life itself is conceptualized (Becker 1994, 386).

One's life course is often conceptualized in terms of a turning wheel, a flowing river, a life journey, or a life span (Holstein and Gubrium 2000, xi).[1] We know about the life course from both formal and informal representations, including religious sermons, medical texts, diaries, government documents, and works of art (Shweder 1998, xii). The life course is also seen as a social institution (Kohli 2007), and acts as an integrating force between individuals and their societies. As a key *temporal institution*, it produces different measures from which one can evaluate one's progression and productivity. From all the above, it can be deduced that the dominant life-course imagery serves as a significant temporal referential frame, through which temporal discontinuities, interruptions, and disruptions are determined.

What is particularly relevant to our analysis here is that the conventional life-course model embodies and defines many social truths concerning individual self-fulfillment, social belonging, and individual movements, predicting in view of these its expected progress and decay. Accordingly, it implies fixed temporal categories and expectations. Consider, for example, this passage written by Esti Avisror, an Israeli single woman, in which she reflects on how as a little girl she used to daydream about her future as a married woman and mother:

> When I was a child, I would flick through the calendar … draw flowers and make calculations. I used to think that by the year 2000—which then seemed so far, and was marked in small numbers—I would be 23 and married, perhaps with a child. I remember, I would close my eyes and wish that I was already there. I believed that this would be the most fulfilling thing that would ever happen to me: I would be a mother. (Avisror 2011)

The above account serves as an example of how the regulatory fantasy of the couple mother-child dyad operates in a context in which the developmental trajectory is fixed and determined. As with many young girls, the writer depicts here how she dreamt of marriage and children, linking these fantasies to ultimate forms of self-fulfillment.

Such examples certify a deterministic life plan with a developmental, linear trajectory, one in which finding a husband and having children are seen as obligatory milestones. According to this view, marriage and parenthood are integral to the life-course progression, and are expected to occur in fixed age cohorts. The concept of life course implies well-recognized categories in the lives of women, and dictates the boundaries of normalcy and sociality. These boundaries are embedded in societal timetables, creating idealized versions of the life course (Roth 1963) and accordingly establish shared expectations and normative judgments. This standpoint emerges from the accounts of many single women. Yael9, a columnist and a single woman, illustrates the interactional dynamic that occurs during family dinners:

> All throughout Passover dinner they were silent … My uncles really held their breath, they were on their best behavior. They were funny; they ate and bragged. Each aunt has a new grandson, each uncle has a secondhand jeep. It was a routine family dinner with many cousins … I was quite surprised, I could not figure out how they were yet to say anything [about my single status] … Have I really managed to train them so well? But then my aunt Esther couldn't hold herself anymore and asked the question that we had been waiting for patiently: Well, you have made a career, seen the world but how about bringing a doctor to the family? (Yael9 2006)

The linear progression, from the family one is born into, to the family one establishes oneself, is perceived as an inevitable future trajectory. Yael9 is expected to follow a heteronormative life narrative, consisting of a required transition from adolescence to adulthood, as well as the obligation to continue the family lineage. Elsewhere in the column, she notes that her career and traveling experiences are now perceived as impediments to the desired procreative life course. This is reminiscent of Diane Negra's observation that "women are depicted as particularly beset by temporal problems that may frequently be resolved through minimization of their ambition and reversion to a more essential femininity" (Negra 2009, 48).

These experiences can be appreciated or tolerated up to a certain point, and only as long as they occur at the right time. Single women above a certain age are a threat to the agreed-upon familial and reproductive life narrative. At this temporal juncture, their presence as unmarried singles sitting at family dinner tables marks a temporal irregularity (Zerubavel 1981), disrupting the expected temporal generational patterns. Pertinent to this discussion is Judith Halberstam's (2005) criticism of what she views as the middle-class logic of reproductive temporality. Halberstam points out that:

> In Western cultures, we chart the emergence of the adult from the dangerous and unruly period of adolescence as a desired process of maturation; and we create longevity as the most desirable future, applaud the pursuit of long life (under any circumstances), and pathologize modes of living that show little or no concern for longevity. Within the life cycle of the Western human subject, long periods of stability are considered to be desirable, and people who live in rapid bursts (drug addicts, for example) are characterized as immature and even dangerous. (Halberstam 2005, 4–5)

Viewed through this perspective, familial life fulfills the pursuit of maturity, continuity, and stability. Complementing Halberstam's analysis, I consider single women to be representatives of a mode of life which defies this customary temporal map, and thus cannot be included with the normative definitions of maturity and civil respectability. The accounts discussed above also reflect the growing temporal awareness of single women. Located in what is grasped as a disruptive temporal stage, their presence draws at this point in time, increasing scrutiny and visibility. The question "When will she marry?" is presented with a sense of urgency, one which expects her to resign to the expected heteronormative timeline.

When will you settle down?

Daniel Levinson's influential model of the life course (Levinson 1978, 57), outlined in *The Seasons of a Man's Life*, is an especially fitting starting point for this analysis. Although it was published at the end of the 1970s, it bears much relevance to current imageries of the life course in Israeli and many other societies. Levinson's theory separates adult development into distinctive and sequential stages, including pre-adulthood, early adulthood, middle adulthood, and late adulthood. In this context, I pay particular attention to what Levinson refers to as the *early adulthood* stage (ages seventeen to forty-five). For Levinson, entrance into the adulthood stage is accompanied by a series of age-related expectations. As such, the thirty-year-old's transition is configured as the period in which one is expected to "settle down," to leave home and to start one's own family.

Indeed, many conventional life-course templates construe this milestone as a critical turning point. In this "settling down" phase, one is expected to "find one's place" and "purpose"; it is a time for realizing one's dreams and potential, a time when crucial choices—namely occupational and marital ones—ought to be made (Levinson 1978). It is pertinent to note here that in recent years, there have been some attempts by psychologically oriented life-course researchers to observe the life trajectories of single women from alternative perspectives. One example can be seen in the work of therapists Natalie Schwartzberg, Kathy Berliner, and Demaris Jacob (1995), who instead of focusing on marriage and childrearing, offer to measure one's life course using benchmarks like work, health, and peer networks. Nonetheless, Reynolds rightly criticizes these models as highly prescriptive, and which consequently do not allow women to define their statuses on their own or to maintain a range of different possible options (Reynolds 2008, 29).

Recent studies of singlehood (for example, Byrne 2003; Moore and Radtke 2015; Reynolds 2008; Taylor 2012; Trimberger 2005) also reveal similar alternative paths available to midlife single women. In some of these studies, being single is framed as an asset; accordingly, women describe their solo life course in terms of individuality and of self-actualization. For many single women (mostly those who belong to the middle- and upper-class strata), singlehood has enabled them to pursue satisfying and successful careers. According to Reynolds (2008), this is a different repertoire within which self-actualization and achievement also gain force, in contrast with marriage. "Financial independence is a goal as well as other more diffuse aims of self-fulfillment, all of which may have been hard won. The notion is that there is so much more to life than getting married or looking after other people, and that without these distractions there is more opportunity to achieve desired goals" (ibid., 60).

Despite the existence of such alternatives, more conservative and developmental life-course models undoubtedly continue to prevail in everyday discourses. Ever-present questions continue to confirm a linear and developmental trajectory: So: "When will you get married?" "Don't you think it's time to be more serious?" "What are your future (marriage) plans?" In Hebrew slang, there is an interesting version of the first question here: *Matai titmasedi?* This can be more or less understood as asking, "When will you settle down?" (i.e. as a legitimate social actor, by participating in the institution of marriage).

The expectation—to be considered as *mesuderet* (settled)—can be found in a letter from a single woman to her mother. The former apologizes, as she knows how much her mother wants to see her *mesuderet* (Bat Chen 2009). The etymological root of this Hebrew word—based on the verb *lehistader*—to become settled or organized—relates to the notion of order, and reflects the importance of creating a stable life structure. In Hebrew slang, it is associated with someone who is well-off or has "made it."[2] Thus, the contemporary slang usage of the word generally denotes economic and personal success; in the context of marriage and family life it refers analogously to someone who is married with children. Indeed, in everyday talk in Israel, one can hear parents discussing whether or not their children are *mesudarim* or *histadru* (the present continuous and past tense of the verb *lehistader*). By the same token, single women often hear the phrase "When will you settle down and bring me *kzat nachat* [some joy]?"

These everyday expressions correspond with prevailing interpretations of what are perceived to be universal life-course models. Social pressures reflect and dictate a structured life course, prescribed in terms of socially defined and carefully timed transitions. Indeed, the settling-down phase itself is not only a period during which one displays one's own potential, but also a period during which the single person is subjected to immense external pressures and societal requirements "to establish oneself," "settle down," and—as in the Israeli slang—*lehistader*. In this sense, "becoming established" signifies not just success and happiness, but also the reaching of a watershed in time and space. The unjustified delays or non-entrance, by the writers above, into early adulthood is in many respects preventing them and their families from joining the desired and orderly life-course trajectory and expected family cycles.

As Halberstam notes (2005), the marking of time according to dictates of marriage and reproduction are connoted to other temporal schemes of investment, insurance, and capital accumulation. Heteronormative common sense, she stresses elsewhere, leads to:

> The equation of success with advancement, capital accumulation, family, ethical conduct, [and] hope. Other subordinate, queer, or counter-hegemonic modes of common sense lead to the association of failure with nonconformity, anti-capitalist practices, non-reproductive life styles, negativity, and critique. (Halberstam 2011, 89)

The Israeli perception of *Lehisteder* reflects Halberstam's observations, illuminating the close links between orderly life-course narratives and normative familial and conjugal discourses. These temporal schemes, as Halberstam notes, are strongly connected. The re-conceptualization of the life course as a cultural and social institution directs our attention to the significant, mostly taken-for-granted normative framework within which perceptions of late singlehood are constituted and maintained. Most of the texts analyzed for this research echo these rigid normative templates constantly, negotiating with the image of life course and its age-related transitioning. In pursuit of this, I turn now to one of the significant factors influencing the socially and historically situated construction of the single life course: postponement of the age of marriage. This analysis does not attempt to align itself with the demographic scholarship often so preoccupied with the delay of the age of marriage, but will rather shed light on the kind of effect it has on the current temporal production of singlehood.

Taking one's time

One of the most significant changes in recent demographic and social trends with regard to intimate relations and family schemas is the delay of the age of marriage. In fact, in many societies the postponement of marriage is often encouraged and is associated with demonstrating choice, individuality, and self-fulfillment. From this perspective, marriage is regarded as a mutual choice and as a contractual agreement between two individuals entering a relationship (Giddens 1992; Swidler 2003); the underlying assumption is that they do so only when they are prepared, at the moment that their relationship has reached the "right stage." These understandings are expressed, for example, in the following excerpt from a *Ynet* column. In the column, the two authors—a psychologist and a family lawyer—outline what they perceive as the three basic requirements for a successful marriage:

- Choose the right person for you.
- Construct the basis of a good and healthy relationship before you decide to get married.
- Nurture the relationship during your time together. (Inhorn and Zimmerman 2007)

These recommendations reflect some widely shared understandings of the "right recipe" for a good and lasting marriage. The second condition—"Construct the basis of a good and healthy relationship"—epitomizes the changing governing norms concerning when and how to prepare for marriage. From this perspective, marriage requires time and preparation. Thus, the delay of marriage is perceived as appropriate; so much so that marrying too young may now be understood as a hasty and less mature mode of behavior.

Early marriage in contemporary secular Israeli society may trigger responses like: "What's the rush?," "You still have plenty of time," or "You still have the rest of your life ahead of you." The two relationship advisors quoted above express their concerns regarding the manner in which many couples shorten or skip over the "necessary" process of building a good relationship. Some of these couples, they explain, decide to get married before they know one another "well enough."

Choice, preparation, nurturing, readiness, and hard work are some of the prevailing buzzwords in the discourse of intimate relations today (Giddens 1992; Illouz 1997; Swidler 2003). The notion of choice is now endowed with more meaning, as it is based on experimentation, maturity, and—most importantly—preparedness. Accordingly, rushing imprudently into matrimony, it appears, is an act that demonstrates a failure to understand and realize what marriage "really entails." It is sometimes even perceived as the explanatory factor in a break up: "They shouldn't have rushed into marriage so fast." At certain phases in one's life, it is of course preferable to wait, to experiment, and to discover oneself. According to this line of logic, achieving economic independence and accumulating life experience is also highly recommended: finishing college; establishing a career; "taking advantage" of what life on one's own can offer. Being on one's own, living alone, experimenting with different relationships, and not committing too early are just some of the recurring cultural recommendations emerging from these beliefs. "Taking one's time" is a required and a respected cultural injunction.

The next extract demonstrates the writer's hesitations regarding marrying too early, alongside her awareness of the social criticism she might encounter.

> I still have time, I know it. I don't want it to happen today. I have no doubt that the age of twenty-one and-a-half is not an ideal age to get married, or to have children. Nonetheless, I already know I want to be there, at the age of twenty-six (after receiving a degree in Communications and being on the verge of a promising career). I want my then-twenty-nine-year-old boyfriend to kneel down and propose. [In my vision], I will agree and will be extremely moved, as I understand that this is not another fantasy wedding plan for the far-off future but a substantial promise for a wedding in the next two months … Many laugh at me when I relay this vision to them. They think I indulge in childish fantasies … but I think it's not childishness or immature, but a biological clock that has started ticking earlier than expected … It seems that we will have to wait patiently for another five years. (Chen 2007)

According to Moran Chen—the writer of this column—her biological clock began to tick earlier than expected. Her wish to get married early is perceived as a childish fantasy. It is interesting to note that when single women delay marriage for too long—as will be discussed later in this chapter, they are also accused of being childish and immature. In another *Ynet* column, a seventeen-year-old single woman who writes to relationship advisers Yael Doron and Gili Bar expresses her hesitation about marrying too young and ending up "like her mother and grandmother, who married their first or second boyfriends and have not experienced anything." On the other hand, she feels that she has found her soul mate, with whom she wants to spend the rest of her life. In response to her question, the two experts explain:

> It is reasonable to give yourself some time. In the meantime, you should live in the moment, as is suitable for your age … And if indeed you continue to feel good with one another as time passes, you could plan the rest of your life together. But in the meantime, give yourselves—both of you—some time. (Doron and Bar 2008)

At this temporal stage of her life, she and her partner are advised to "give themselves time." As they are still considered young, *time is still in their possession.* The complexity of the demographic and social transitions leading to these social dilemmas is also illustrated in another one of Yael and Gili's advice columns, which depicts this changing social reality:

> Many singles are still not coupled for many different reasons. There are those whose order of priorities have led them first to study, to create a secure economic basis, to be independent; and only now are they ready for coupledom. Others had to sort things out for themselves before they were able to turn to committing to long-term relationships, whether it was due to their family of origin, their childhood, or other "baggage" they carried with them. The rest are … late bloomers; at the age of thirty-three "they've woken up" and have now reached the stage at which their cohorts long ago found coupledom and settled down. (Doron and Bar 2009)

After purportedly illustrating the diversity and complexity of the reasons behind the varied entry points into marriage, the advice column ends, nonetheless, wishing for its readers in bold letters: "May the next date be your first and last!" (ibid.).

Hence, according to Yael and Gili, the rise of single households is a result of a variety of factors. As always, the reasons for singlehood must be deciphered and explained. Yael and Gili rationalize what is considered to be a social aberrance. They point out that some single persons have had to take care of their professional careers and economic independence first, while others were not ready emotionally, or are late bloomers. In other words, late singlehood always has a reason which has to deciphered and accounted for. Nonetheless, the overarching refrain remains: "May the next date be your first and last!"

Speeding up

> You tell yourself that it's ok, you feel wonderful, and you are not pressured yet, but deep inside you know that each man that passes by causes you to consider the possibility of coupledom with him … meanwhile the clock is ticking. They say that above the age of thirty, girls become hysterical from the pressure to get married. (Or-Li 2008)

Tamar Or-Li, the single Israeli woman quoted here, describes how her temporal awareness changes, and refers to the well-known metaphor of the biological clock. At this point in her life, she begins to sense that her time is running out. The social message is clear-cut: one should quickly hop on the next train and join the ride before moving from the category of a "late bloomer" to that of the "old maid." As such, I propose that the postponement of marriage is still very much limited by conventional socio-temporal regulations. While getting married at seventeen—or twenty-one for that matter—to a first or second boyfriend has come to signify a premature social move (at this age, one has all the time in the world, and living in a meantime or liminal mode is encouraged), a few years later single women must bear the social responsibility for ignoring their ticking biological clocks.

Consider, for example, the stereotype of the choosy or selective single woman (Lahad, 2013). While in the earlier stages of singlehood her self-determination can be admired and praised, upon reaching what society considers to be a suitable marriageable age—and particularly after passing this—the choosy single woman is perceived as trapped within her own self-regulation. She is the sole saboteur of her future happiness and well-being. At this point in time, her selectiveness marks an overstepping of socio-temporal boundaries, and a disruption of the expected and socially mandated life schedules.[3]

The transition is vivid. Time should be accounted for, should be managed carefully; when a woman is at the prime age for marriage, she should hurry up. At this time, single women are warned again and again that they have no time, or that they are running out of time. At this life stage, time is a sacred and limited resource and accordingly they *no longer* have the luxury of occupying the "meantime" position and delaying marriage. It turns out that the newly articulated required delay becomes, in a few short years, socially intolerable and unjustifiable.

Merav Resnik, a columnist writing on *Ynet* explains:

> If we don't play the game of life, the game will continue without us. And then one day we will decide to get up from the bench, look around, and wonder how the hell we got here. Sometimes we decide to get up one second before the game is over so that perhaps something could be salvaged from this mess. Sometimes we wake up too late, which brings us to the platform of missed trains ... "Too little, too late" is the slogan of this platform. Those on the platform are standing, waiting, fearful, expecting the train that was here and is now gone ... We [single women] are left on the platform alone. The one who was supposed to hop on with us has boarded the train by himself or perhaps has found someone else to go with. (Resnik 2007a)

Delaying marriage for too long cannot be justified, because at some point one simply "misses the train" for good. As I noted in the Introduction, the train metaphor is abundant in discussions of singlehood. In these accounts the train represents an agreed-upon collective orderly movement, one which reiterates the heteronormative temporal trajectory. One can make it just in time, get on at the right station, join the train, and move ahead with the others, miss the train or take the wrong one. As their singlehood proceeds, the chances of hopping on the right train become minimized. As

Merav observes cynically, there is a special platform for older single women: "the platform of missed trains." These shared cultural assumptions highlight how time and the socially constructed notions of the life-course are standardized. The hegemonic paradigm of the life course, epitomized in this case by the train, is a reference point against which rigid standards are set. Drawing on the train metaphor highlights the way single women position themselves in relation to dominant life-course models and their dictated rhythms.

The encouraged delay is now re-conceptualized, and regards the "aging single woman" as facing the imminent danger of being left alone on the now deserted plat-form. Her time slot is squeezed, and it may be deduced that the delay which allowed more time, now poses the risk of leaving no time left at all. Actually, the cultural legiti-macy granted to the postponement of marriage substantially *shortens* the amount of time single women have to find an "eligible partner." Thus, the newfound legitimacy granted to postponing marriage does not, by any means, indicate that single women's time slot is an unlimited one. The years during which they are still "eligible" and "trad-able" are just as restricted, yet in a slightly different temporal frame.

What was perhaps viewed as readiness, maturity, and experience at the age of twenty-six, for example, has been drained of its prior meanings and is now consid-ered as an obstacle, indicative of developmental incompetence. For example, the cul-tural injunction to experience being on one's own, to experience independence, is now replaced by the warning that one will "end up dying alone." As noted above, after a certain period of time, single women can be considered as too independ-ent, too selective, or too lazy (Lahad 2013). In that respect, they are at risk of *losing time*, therein losing their agency. At this stage, they can no longer *take their time*. Their age and single status becomes their master status (Becker 2008), and in this respect their deviance from societal timetables could be labelled as "pariah feminini-ties" (Schippers 2007). The refusal to abide by these temporal schemas could poten-tially contaminate gender relations between men and women. As Schipper explains, a woman exhibiting defiance, physical violence, or authority in a patriarchy can potentially destabilize male dominance, unless the exhibit can be stigmatized and feminized.

This shift in interpretation demonstrates the extent to which personal qualities and behavior are evaluated within these age-based parameters. This now highly condensed timeframe often results in growing pressures, and leads to *time panic* ambiance. Stan-ford Lyman and Marvin Scott have defined this type of panic as:

> Produced when an individual or a group senses it is coming to an end of a track without having completed the activities or having gained the benefits associated with it or when a routinized spatio-temporal activity set is abruptly brought to imminent closure before it is normally scheduled to end. (Lyman and Scott 1989, 46)

Or, as Negra poignantly claims, one of the characteristics of postfeminist culture "is its ability to define various female life stages within the parameters of 'time panic'" (Negra 2009). In the case of "late singlehood," the routinized temporal patterns addressed here produce and enforce an unequivocal social message: single women above a certain age are entering what Reynolds (2008) has termed as the *twilight zone*. While scanning some of the columns from *Ynet* discussed earlier, I came across a very telling *JDate*[4] advertisement announcing: "Only When You Are Ready: JDate." This message corresponds with the life-course developmental models, and can be grasped as the settling down stage. The advertisement invites the male surfer to find single women between the ages of twenty-five and thirty-four: this temporal framing sheds light on the age-graded system through which singlehood is socially interpreted. This line of analysis will be explored in detail in Chapter 4.

The rapid and confusing social changes of recent decades are reflected in this dialectic. On one hand, viewing singlehood from the socially constructed life-course perspective reflects the dynamic nature of age-related assumptions. On the other hand, embracing a nuanced look at singlehood also reveals that our perceptions of the life course are still very much stage-ordered, implying temporal regular patterns even as they introduce ever changing age-stratification norms.

So far, this chapter has covered some of the striking aspects of the normative life-course model. I now turn to discuss what I view as another significant parameter within the traditional life-course models, the belief in developmental and linear progression.

The linear trajectory

The centrality of the linear logic is demonstrated by the conventional thinking about "late singlehood." Single women are constantly asked, explicitly and implicitly: "What's going on?" "What's new?" "Any news?" An abundance of visual images depict single women as waiting for a telephone call, for a sign, or for "Mr. Right." Warnings, such as "In the end you will die alone," and blessings and consolations like "By your wedding day you will feel better," suppose a continuous and coherent linear life trajectory. I argue that the comforting tone adopted in a phrase addressed to an upset childlike is embedded into a heteronormative destiny, one which determines a linear path that will certainly lead one to one's wedding day. Getting married captures the promise and the pervasive expectation that at a certain point everything will "work out," and that the temporary disrupted order will accordingly rearrange itself.

Moreover, these blessings signal crucial signposts which structure and bestow meaning upon the measurements and movements of time. From this standpoint, when one wishes a single woman: "*Bekarov ezlech!*" (Soon at yours [wedding]!), this expectancy forms part of a normative injunction emphasizing a linear developmental order. In Chapter 7, I analyze in greater detail the ways in which this blessing marks a climb up a *linear temporal ladder*, wherein single women can join the social order once they are married. My contention is that, according to this logic, the single woman is located in a symbolic *heteronormative queue*, within which all she can hope for is to move forward and shorten her wait to the minimum. Here, I wish to stress that this blessing reflects the manner in which our social life is constantly organized and regulated by temporal linear indicators, as well as by the ways in which participation in this pathway is highly encouraged.

In Julius Roth's study of the temporal experiences of patients in hospital, Roth was particularly interested in how the structure of time imposes certainty and predictability upon the trajectories of hospitalized patients:

> One way to structure uncertainty is to structure the time period through which uncertain events occur. Such a structure must usually be developed from information gathered from the experience of others who have gone or are going through the same series of events. As a result of comparisons, norms develop for the entire group about when certain events may be expected to occur. When many people go through the same series of events we speak of this as a career, and of the sequence and timing of events as a career timetable. (Roth 1963, 136)

Roth's evaluation can be applied to the *Bekarov ezlech* blessing or to reoccurring statements like "By your wedding day you will feel better." The social knowledge expressed here takes the form of collective benchmarks and sign posts. In this way, information is gathered from the experience of others, constant comparison seeming to hold the entire group together. In a similar vein, the career timetable of the Israeli single woman is prescribed in advance, and social injunctions therefore spur her to move forward in a pre-defined and recognized linear trajectory. The linear trajectory of a woman's life course is utterly predictable.

Linear time's most powerful claim, observes Carol Greenhouse (1996), lies in its own redemptive power in relation to individual life. This hope and belief is expressed, for instance, in the well-worn cliché, "Don't worry, eventually you will find true love." Not only does it presuppose a long linear journey, but it also offers individual redemption and salvation. Queries, such as "What are you waiting for? "What's new?" are set within this linear logic. Heteronormative and patriarchal assumptions structure time, and in this sense imply that all that the single woman can hope for is to "move forward in life" and find the right guy to marry.

The linear homogenous time also assumes a linear causality model providing one with a utopian *telos*. In contrast, this mode of explanation—"In the end you will die alone"—echoes the reasoning of a linear trajectory, one which makes an indisputable connection between the present and the future. For an unmarried woman, her future holds nothing but misery and loneliness. Thus this statement can be interpreted as the single woman's diminishing agentic ability to *determine her future*. As a consequence, these dreary forecasts acquire deterministic-linear and even fatalistic undertones. Long-term singlehood represents a non-progressive, non-developmental linear path. The image of the train is especially relevant here. It accentuates not only the importance of being on time, but also the sequential linear track composed of a chronology of life stations determined by basis of age.

This formulation also fits well with what Anthony Giddens (1990, 1991) views as modernity's obsession with the *colonization of the future*. Embracing such an approach entails the promise of attaining greater mastery of one's life trajectory and bestowing one with certainty and predictability. Within this dominant framework, heteronormative expectations constitute singlehood as "being on the way": the journey might take some time, yet the desired destination is certain. In a similar fashion, singlehood also feeds into what Zygmunt Bauman (1996) has termed a modern individualistic pilgrimage. Bauman writes that unlike the pilgrimages of pre-modern times, the pilgrimage of modern individuals is accomplished without them actually leaving the home. In modern society, this journey is not a choice but a mode of life. Hence, the world of the pilgrim is orderly, predictable, assured, and progressive:

> For pilgrims through time, the truth is elsewhere; the true place is always some distance, some time away. Wherever the pilgrim may be now, it is not where he ought to be, and not where he dreams of being. The distance between the true world and this world here and now is made of the mismatch between what is to be achieved and what has been. The glory and gravity of the future destination debases the present, plays down its significance and makes light of it. For the pilgrim, what purpose may the city serve? For the pilgrim, only streets make sense, not the houses—houses tempt the tired wanderer to rest and relax, to forget about the destination or to postpone it indefinitely. Even the streets, though, may prove to be obstacles rather than help, traps rather than thoroughfares. They may misguide, divert from the straight path, lead astray. (ibid., 20)

Bauman's conceptualization of the spirit of pilgrimage expands our understanding in relation to some of the common representations of the existential experiences single women might expect. The liminal position of singlehood is likewise configured as always being "somewhere else," and the emphasis placed on the future destination reminds us all that the single woman is where she is not supposed to be, both spatially and temporally.

Prevailing images of single women today are still formulated and constituted in much the same way as the modern pilgrim metaphor. The roads to "Mr. Right" are the streets through which the single woman should march. There are no houses in which she can rest and forget about her future destination. From this perspective, Ulrich Beck and Elizabeth Beck-Gernsheim have underscored how the pressures to construct one's life as a biographical project or *do-it-yourself biography* stress the contemporary injunction for individual life planning (Beck and Beck-Gernsheim 2002, 3). This line of analysis will be developed further in the coming chapters, but for our purposes now it suffices to note the clear connection between individuality and linear time. Individual self-identity, as Ÿian (2004) puts it, is reflected through images of the future self. For Ÿian, the future-oriented mode of constructing identity is performed by universalizing and individualizing aspects of linear time in terms of conceived uniqueness and autonomy:

> In general self-identity tends to be mirrored in images of oneself in the future. This is a future that is understood as open space to be realized by individuals. Such a conception of the future is provided by linear time in cognitive and ideological categories where individualism and linear time appear to be two sides of the same coin ... linear time cannot be properly understood unless seen in relation to individualism. Firstly, the individuation of people that historically has taken place through citizenship, wage-work, money and commodities can hardly be seen independently of chronometric instruments of linear time (the clock and the calendar). Secondly, it is by highlighting our wishes and ambitions for the open future that we can escape the constraints of the present on the

autonomy that is preconditioning the experience of our own individuality as persons. (ibid., 175)

In his work on the temporal experiences of the unemployed, Ÿian contends that the unemployed lack the basic qualifications that would enable them to participate in the collective linear temporal orientation; therein, they experience no sense of certainty or progression from the present to the future. A similar claim could be made in relation to widespread beliefs about "late singlehood." By not joining the collective linear path, single women are commonly perceived as lacking the basic properties which would in turn enable them to join the collective movement forward.

In this context and before concluding, it is important to note that linear visions of the future are interwoven with cyclical ones. Cyclical temporal idioms permeate conceptions of private and personal life. Birth and death, the rise and fall of generations, the transformation from dust back to dust, marriage and parenting, all follow cyclical rhythms (Daly 1996; Greenhouse 1996, 23). From this outlook, the construction of family life exhibits many cyclical qualities. For instance, when one wishes for the single woman to get married soon, linear conceptions are employed to attain familial continuity. Thus, the concurrent search for continuity, repetition, progress, and change in the normative construction of family life is expressed in both linear and cyclical temporal formations. The power of these temporal forms resides in the constant pressure to move forward within the linear trajectory, and to reintegrate into society by following both linear and cyclical rhythms. Lifelong singlehood, then, is configured as an aberration and a distraction from appropriate modes of linear and cyclical temporality. Single women are identified as the living representation of this irregularity: they are perceived as wasting time, putting their lives on hold, or living in empty time. In the next chapter, I will try to further understand the ways in which this disruption is fabricated and enhanced.

Notes

1   See also Becker (1994).
2   For example in 2007, Israeli TV broadcasted a sitcom called *Mesudarim*, about four Israeli friends who "made it" by selling their high-tech start-up company and becoming millionaires. The term could also be read in gendered terms. In Israel the term is generally used to refer women who have succeeded in getting married, usually alongside having had their own property and a steady job. From this perspective, single women in Israel cannot fit the *mesuderet* rubric.
3   See my discussion of single women's selectiveness, Lahad (2013).
4   A popular internet dating site.

3

Singlehood as an unscheduled status passage

In the Introduction, I wrote about an encounter with a colleague, during which she asked me how it was that a woman like me was still single. The presumption must be that this question wouldn't have been asked if I had been in my early twenties or late forties. But as a thirty-year-old woman at that time, her question was infused with a sense of urgency, and the hope that my single status would soon transform into a married one. It is worth paying attention to the expected temporal status passages that underpin interactions such as this. Precisely because these matters are rarely problematized, they warrant closer attention. In this chapter, I explore this transition as a social construct, and explore the various ways by which the status is produced and maintained through situated social norms and regulated temporal codes. To expand this analysis, I draw on sociological and anthropological studies, as well as considering my findings in relation to symbolic interaction traditions.

This chapter expands the analysis of the expected linear life-course trajectory as discussed in the previous chapter, but from a different perspective. The focus here is a conceptual analysis of *becoming single*, through which I explore the discursive mechanisms and regulations that generally shape it as a biographical disruption. In the chapter, I argue that this process is undertheorized in relation to singlehood, and its temporal assumptions rarely critically examined. Timely status transitions are configured as part of a default life trajectory subjected to socio-temporal schedules. As Zerubavel (1981) stresses, schedules are responsible for the establishment and maintenance of temporal regularity in our daily lives. It is these temporal regularities that I wish to observe, casting a critical light on how and why they are taken for granted.

My discussion examines this path in relational terms, through which the process of becoming single and the transition from "normative" to "late" singlehood is produced by socio-temporal truth statements. Thus, the stages of singlehood—or more specifically what I term as *singlehood career*, drawing on Goffman's use of the term—come into existence through a hegemonic temporal gaze. Throughout this chapter, I show how this gaze is established through social interaction, and is deeply ingrained in collective socio-temporal perceptions. Following this line of inquiry, I consider how the discursive switch of becoming single operates as a subtle non-institutionalized transition process, one through which the entry and exit from "normative single-hood" to "late singlehood" occurs without rituals or official formalities to accompany the change.

The second part of this chapter offers a temporal reading of the question "Why is she still single?," a question that many single women repeatedly hear and ask of themselves. My intent here is to explain the discursive formations and implications of this ubiquitous question, and to shed light on how popular knowledge about single women is produced and circulated. Thus, instead of asking why single women are single, I study how the inquiry is discursively constructed in relation to how singlehood, as I term it, is established. I view this process as an interpellative process, following Louis Althusser (1971), in which the woman as a subject is urged to respond. When single women ask themselves this question, they internalize the cultural stereotype that exists of themselves, as single women. However, as will be shown, single women find ways to subvert this question and the power of this dominant discourse.

Charting the single life-course

I begin my analysis with Louise (a pseudonym), regular contributor to *Ynet*, who outlines the following definition of the single-life-course trajectory:

18–23: Too young to be looking for something serious.
23–27: She is just too successful.
27–29: She is too picky, she will end up alone.
29–32: She is just too lazy. "Would it kill her to go out on the blind date her father set
　　up? So what if he doesn't know the guy, he knows his parents. It is a good family."
32–35: She has given up. What a pity. She used to be so beautiful. (Louise 2007)

This diagram is an excellent departure point for our exploration of how perceptions about single women's life trajectories cannot be isolated from contemporary temporal orders. In that sense, single women's temporal locations are not fixed or static categories, but rather ever-changing. It is interesting to observe the social mechanisms through which the reasoning of singlehood is subsumed. Thus, the proposed line of inquiry here highlights the socially situated aspects of the single woman's life trajectory. Louise demonstrates that under this collective temporal gaze, single women are objectified according to age-based evaluations: they fluctuate from being too young to becoming too old, picky, or lazy, until they cease to be suitably competent to participate in the search for a male partner.

The different life stages illustrate the elasticity of social definitions, and how social boundaries and categories are constantly created and subjected to arbitrary changes. Accordingly, each temporal phase determines the attributes and market value of single women. For example, between the ages of twenty-three and twenty-seven they are considered as "too successful," while between twenty-seven and twenty-nine they are "too picky." In this manner, Louise provides insightful commentary on how social mechanisms invent and configure social categories and boundaries. These expectations turn into constant self-surveillance, through which single women are expected to carefully monitor their life-course phases and police themselves according to the ensuing expectations.

One way to understand the discursive construction of the single woman's life trajectory is through the term *status passage*, coined by Barney Glaser and Anselm Strauss (1971). Their study differentiated between charted, planned, regularized passages and emergent passages, which are open-ended and are constituted as they occur. My claim is that this kind of differentiation can present a rich observational perspective regarding the changes in relation to single women's social status along the life course.

For this reason, I find Glaser and Strauss's analytical framework particularly useful, as it assists us in exploring the micro-status transitions which are over-simplified and rarely problematized. A more nuanced understanding of these transitions can reveal more about how they are determined by the disciplinary gaze of others; a critique of this gaze can present new opportunities for imagining alternative, more heterogeneous pathways. I wish to argue that one of the challenges of theorizing singlehood is the *unarticulated entry and exit rites* that define it. Unlike conventional perceptions of marriage and parenthood, there is a less agreed upon, visible, and tangible entry point to the stage at which midlife or late singlehood begins. As opposed to, for example, the still increasingly popular bachelor and bachelorette parties—which designate a clear exit and passage designating movement from one status to another—no such rituals or ceremonies are available for single women.[1] Drawing on the work of Glaser and Strauss I term late singlehood as a *non-scheduled status passage*, as late singlehood has no institutional schedules of its own. Building on Zerubavel's (1981) formulation, this represents a temporal irregularity.

An example of a disruption of these expectations can be found in a *Ynet* column by Harela Yishi (2013) entitled "From the Queen of the Class to a Forty-Year-Old Single Woman: How Did It Happen?" Yishi, a relationship advisor, claims that at a certain phase in their lives, all single women who believed that "they had it all" end up on their own, alone. The illustration chosen to accompany the column depicted a woman sitting by herself on a beach with an empty chair next to her. The image was accompanied by a telling quote: "I thought I had all the time in the world, but then I remained alone." In common with Louise (2007), the single woman thought she had all the time in the world but lost her agentic capacity for time. The image of the solitary woman may also signify this involuntary "progression," or some kind of *downward mobility* associated with the image of the lonely single woman. Moreover, the question "How did this happen?" also marks a disruption to the sense-making mechanisms which are also embedded in the "Why is she single?" query, which I will return to in the second part of this chapter.

Helen Ebaugh's work (1988) on role exits draws upon Glaser and Strauss's research, and suggests that emergent passages are characterized by fewer precedent guidelines to facilitate the passage. Ebaugh also differentiates between role exits which are clearly defined by society—such as the retired worker or medical student—and non-planned passages—which are far less institutionalized—such as nuns who leave their religious orders (ibid., 155). Building on these lines of inquiry, I suggest that long-term singlehood could also be understood as a non-scheduled status passage. My point here is that exiting the normative singlehood phase and entering the late singlehood phase often lacks the structured expectations, rites of passage, and institutionalized socialization processes that are associated with moving in together, getting married, or having children, to state a few examples. It does not have a discernible time-related benchmark, nor is it identified by practices, symbols, or objects.

It might be that the expectation is that this elusive, yet vivid formulation would provide the answer to a question I have often encountered during my research on single women: *When exactly does late singlehood begin*? Friends, colleagues, students, and strangers have repeatedly posed this question, as though the question can be answered definitively. Might it be twenty-six, thirty, thirty-two, thirty-five? The answer varies, obviously, from one single to another, yet it is possible to sketch a more or less social timeline according to which the transition can be expected to take place. This could be further complicated by looking at these points of transitions as the different phases of *the single woman's career*, to borrow Goffman's (1961) well-known formulation. Goffman developed the concept of career to explain social transitions in everyday life and not only in occupational terms.

In Goffman's celebrated work on asylums, he referred to the moral career of patients as the progression of individuals through a number of social roles that position them as mental patients, as well as the way in which a circle of agents participate in the individual's transition from a *civilian* to a *patient* (ibid.).[2] The aforementioned column (Louise 2007), when considered in light of Goffman's observations, draws our attention to the various ways in which female singlehood is interactional and socially accomplished. The sort of analysis suggested here understands singlehood as a process that is made and negotiated, and not as a stable social entity. In it, meanings of singlehood can be varied by differentiating the features that are themselves socially constructed and altered; in this specific case, differentiating the age cohorts and life stages of single women. Recognizing the constructed nature of the life-course highlights how singlehood is mediated by the meanings we attach to different life-course phases, and their expected transitions.

In between status lost and status gained

In her study on divorce, Nicky Hart (1976) offers a nuanced look at the process of marital breakdown. Hart argues that during the process of divorce, there is no precise point at which a person feels the complete disassociation from one's former life partner.

> Being divorced is better thought of as becoming something than as occupying a fixed social position with well demarcated boundaries in time … Passage from one status to another is rarely a clear-cut process. The status change may be marked by a ritual act and by an abrupt change of title or behavior. But in most cases the boundary between status lost and status gained is blurred; not only is the new status often partially experienced before the public rites of passage but also the life associated with the old status may never be completely discarded. (ibid., 103–104)

In contrast to marriage, the entry point or transition to late singlehood in some ways resembles Hart's perception of divorce, in that the transitions in the single woman's life course seem more subtle and less concrete. In that way, the boundaries are less well-defined, and the passage indeed is not a clear cut process. Still, in common with other religions and civic cultures, divorce in Judaism is ritualized and replete with sets of temporal symbols which accordingly mark time and bestow it with meanings and values.

Against this, I contend that the change of status—namely transitioning from a "young single woman," or being associated with what is perceived as normative single-hood phase, to becoming "a woman who has passed her prime"—is not an abrupt, straightforward transition but rather an *unstructured and unscheduled* one. It can be perceived as a private, invisible and subtle process, but at the same time it is highly visible. The blurred temporal boundaries that result can be one of the reasons why entering the late singlehood phase leads to so much confusion and heightened anxiety. The process of becoming a single woman and the different unstructured phases of singlehood lack the sense-making devices, role expectations, and orientation rules which formal rituals supply.

This dynamic was beautifully exemplified in a conversation I had some years ago with my cousin, who was eight years old at the time. While discussing the advantages and disadvantages of single life, she asked whether her eighteen-year-old sister could be considered to be a single woman. Her question poignantly illuminated the subtle process by which one becomes single. It is perhaps the case, then, that singlehood becomes a tangible and visible category only when marriage and family life appear on the life-course trajectory as a concrete option. One is much less likely to ask a sixteen-year-old girl if she is single or married than a thirty-five-year-old woman.

Hazan observed a similar lack of rites of passage at a day center for elderly people in London:

> There is no element of preparing the participants to enter a new phase in their lives ... There are no official instructors, no official novices, no recognized stages through which participants pass, and no social recognition of the conversion. (Hazan 1980, 147)

In some respects, the same sentiment could be applied to single women. There are no preparatory processes, no official instructors, and no novices. In everyday talk, we often refer to the idea of *becoming single* after experiencing a visible transition point such as separation, divorce, or widowhood. Self-help books like *Becoming Single: How to Survive When a Relationship Ends* (Keith and Bradley 1991) stress how divorce can operate as a de-coupling process, marking a new social phase in one's life trajectory. This kind of guidance and advice, as offered by such self-help books, seeks to help individuals negotiate this transition.

The unexpected change of status in the case of late singlehood requires different sorts of social learning and heightened reflexivity. For example, Lalli Blue, a single woman and *Ynet* columnist, observes: "My status as a single woman is a total fact. I have learned to look into the glass and to see the wine, to see the good things" (Blue 2006). This passage acknowledges that the transition into a late single woman involves a process of social learning. Unlike marriage and family life, social life does not prepare one for late singlehood; there is no recognized or institutionalized process of socialization for such a status. No one hears of an adult preparing a child "for the rigors and requirements of late singlehood," primarily because this stage is not granted the social legitimacy accorded to other statuses. Indeed, the very idea sounds absurd. It is not a given self-evident process, but rather demands personal and social re-adjustments, both by single women and their environments, and a *temporal re-orientation* is one of these. The need to readjust stems also from the absence of positive models for long-term singlehood, and no preparation or socialization for this life path.

The desire of single women to get married within a few years or very soon dictates planned future pathways. Under these discursive conditions, wishing for a young man or woman to stay single is not an option; it might be interpreted as an insult or even a curse. The "old maid" cannot serve as a role model (and there are hardly any "positive" options that involve lifelong singlehood); consequently, the possibility that one might become such a relic is not even discussed.

The common understanding of singlehood is of a transitory and preparatory phase. During this time, one ought to prepare for married life. This unfeasibility of discursively *structuring* the single status further accentuates its prescribed emptiness, incoherence, and lack of meaning. My contention here is that there are strong links between the absence of positive models of late singlehood, and the lack of a structured transitioning process to the configuration of late singlehood as a disruption.

"Late singlehood" as a disruption

The idea of "late singlehood" disrupts the cultural expectations about life-course schedules. An analysis of the various texts discussed thus far illustrates the manner in which this disruption is experienced, and highlights especially the gaps between cultural norms and the subjective experiences of single women. The term "late singlehood" in itself represents a disruption and, as such, a disparity. In other words it commonly represents an unjustified delay which "has gone too far." In certain ways, the unstructured transition into lifelong singlehood indicates a separation from collective timetables and societal rhythms without re-integration to new ones. What emerges from many of the texts examined in this book is that many single women, from a certain point in time, are perceived as disoriented, marginal subjects.

This experience of temporal a-synchronization is one of the reoccurring experiences among many single women. It is manifested, for example in the questions that single women are asked again and again, e.g. "When are you going to get married?" In the texts analyzed throughout this study, marriage and family life are still very much considered as the natural, required steps that align with the dominant norms that signify maturity and success. This vision is reflected in the next passage:

> When we were young, the only reason to fear the holiday dinner was the food ... The same dinner for single women who have passed the age of twenty-seven can be experienced as a military exercise in jumping through flaming hoops ... In a world where marriage is considered to be worthy of a Nobel Prize ... the single daughter, the single

grandchild, and the single niece are considered to be failures while all of the other women in the family hold the prize [of marriage]. (Thelma and Louise 2006a[3])

For the writers in question, the point of transition comes at the age of twenty-seven. A clear hierarchy is erected, once again, between the singles and non-singles sharing the same dinner table. Here, I wish to argue that late singlehood is not granted with the symbolic privileges and profits associated with the nuclear model. As Pierre Bourdieu notes, the family is:

> a de facto privilege that implies a symbolic privilege—the privilege of being comme il faut, conforming to the norm and therefore enjoying a symbolic profit of normality ... The family plays a decisive role in the maintenance of the social order, through social as well as biological reproduction, i.e. reproduction of the structure of the social space and social relations. (Bourdieu 1993, 23)

Bourdieu's observation sheds light on what bestows this hierarchy with so much power and logic. My formulation of singlehood as an aberration to this normative temporal ordering is also indebted to Gay Becker's (1994) important scholarship on disrupted lives. Becker sees disruption as a multifaceted cultural process, one which enables re-examination of the disparity between cultural ideals "of how things are and how they actually are" (ibid., 401). As she explains elsewhere:

> When expectations about the life course are not met, people experience inner chaos and disruption. Such disruptions represent loss of the future. Restoring order to life necessitates reworking the understanding of the self and the world, redefining the disruption and life itself. (Becker 1999, 4)

In a study of youth unemployment in Norway, Hogne Ÿian follows Gaston Bachelard's argument by stressing that lived time is not rooted in time as an *a priori* form or substance, because time is not an integrated or a one-dimensional continuity, but rather is discontinuous (Bachelard 2000 cited in Ÿian 2004, 181). In a study concerning unemployment in Austria, Marie Jahoda and Hans Zeisel (1974) contended, for instance, that leisure for the unemployed was a tragic gift, as for the unemployed the division of the day into hours had lost all meaning

Hence, the notion of continuity and sequence are a primary meaning-making device, sustaining the illusion of order against the backdrop of inherent disorder (Becker 1999). This point is also developed in Becker's study:

> We fool ourselves into thinking that the world is an ordered place. That's how we get up in the morning and how we go to bed at night, because we are ordering the world in some fashion ... and it is just an illusion, an illusion that keeps us going. If we didn't pretend that that's the way it is we wouldn't be able to function. (ibid., 63)

Amongst the common threads linking these experiences are the complex temporal negotiations people undertake when they are distanced from the linear time path. This separation and even disorientation demonstrates a distancing from society's main institutions and sense-making mechanisms. Marginalized groups, such as the unemployed, prisoners, the homeless, or chronically ill persons, are symbolically removed from partaking in collective linear movements, as they "fail" to keep up with normative schedules.

Gerda Reith's (1999) research about the subjective aspects of time among drug addicts also shows how linear perceptions of development and progress are no longer relevant for many of her interviewees, which leads them into creating their own time. In various other studies on prisoners (Scarce 2002), the chronically ill (Charmaz 1997), and HIV patients (Davies 1997), researchers have illustrated the manner in which groups construe their own timeframes. Michele Davies, for example has attempted to understand how the diagnosis of and living with HIV among HIV patients has changed their orientation towards time and the ways in which they seek to compensate for the loss of the temporal assumptions that existed prior to the diagnosis. Davies has found that when a person receives a diagnosis of HIV, they are immediately taken aback by the certainty of the assumed futurity of their existence. Becker argues poignantly that when things are synchronized, time can go unnoticed, yet when it is disrupted it becomes visible.

These varied perspectives provide rich insight into how hegemonic perceptions of time establish normativity and stability, and accordingly produce "late singlehood" as an abnormality and a mistake to be fixed. This normative temporal discourse on singlehood devalues any disruption to the linear developmental life trajectory. This particular temporal experience can be seen as a biographical disruption reiterating Michael Bury's (1982) term. In his analysis of chronic illnesses, Bury analyzes how they disrupt daily lives and the knowledge associated with it. According to Bury, one is abruptly acquainted with experiences of suffering, pain, and death, which were previously seen as distant possibilities. His articulation of the idea of distant possibilities is also highly relevant for the cultural production of "late" singlehood. When the prevailing norm is that "By your wedding day you will feel better" or "Soon at yours [wedding]!," life long singlehood is rarely expected. As Hazan notes:

> The social patterning of time, which originates in the inability to conceptualize a continuous flow of change, may take various forms. Cultural codes breaking up time into symbolically recognizable units serve to make sense of experience. When these codes lose their social validity or cease to reflect experience, temporal construction collapses. (Hazan 1994, 74)

Lifelong singlehood is not a possibility which can be imagined through the socialization process, and is not indicated in life-course charts and diagrams. However, as I will further elaborate in the last chapter of this book, there are new voices which present long-term singlehood as a viable and non-disruptive option, and accordingly view this option in terms of self-development and achievement. For example, a recent study conducted by two Canadian psychologists (Moore and Radtke 2015) found that some of the midlife single women they interviewed constructed their midlife as a time of transition, one in which, "following a period of critical self-examination, re-evaluation, and action-taking—one aims to create a stable, economically secure, and satisfying life as a single woman" (ibid., 310). Moore and Radtke have termed this subject position "comfortably single at midlife women," as these women have accepted their single position as an independent and viable way of life.

This finding stands in contrast to deficit subject position (Reynolds 2008), which emerges from the above mentioned Israeli columns in which the status passage to late singlehood requires a reassessment of one's position, as familiar sense-making devices are disrupted. This breakdown in the articulation of time is revealed and validated by some of the most disturbing sets of questions encountered by most late single women: Why are you still single? What's stopping you from getting married? However, instead of responding to this set of questions and accepting its heteronormative logic, I propose to explore the way it reflects temporal schemes and normative life trajectories.

Why are you *still* single?

> Let's take for example Ella, a thirty-four-year-old woman ... beautiful, witty, intelligent, and funny. Apparently no one can understand why she is alone. (Brodsky-Kauffman 2008a)

> Ayelet ... is a thirty-four-year-old woman, a beautiful and attractive woman. She has an MBA and holds a senior position in a big company. She belongs to a middle-class family but she has never been in a long-term relationship ... She knows how to flirt; she keeps her eyes wide open and recognizes what she has in front of her. Nonetheless, she struggles to understand how it can be that despite all of her wonderful features, she is still alone. (Brodsky-Kauffman 2007a)

The excerpts above were penned by Esta Brodsky-Kauffman, a dating advisor. Many of Esta's columns, published on the Israeli portal *nrg*, deal with what she refers to as a new social problem: the thirty-something, good looking, intelligent, successful, yet still single woman. In this part of this chapter I offer a temporal reading of this question so prevalent in discursive patterns about single women. My intent here is to comprehend its discursive formations and to discern how these regimes of truth exclude other forms of knowledge.

Most commonly, this question is expressed with more than a touch of surprise. It first sets out to understand "what went wrong," and secondly seeks to uncover the hidden reason(s) for one's late singlehood. This question is often raised by singles themselves: "Why am I single; what is wrong with me?" These questions and self-doubts emerge as pervasive disciplinary apparatuses (Foucault 1991), that can be asked by anyone at any time. Sociologist Margaret Adams proposes turning the tables:

> Confront a happily married woman with the same question: what happened, Martha, Eleanor, Patricia ... that you got married? ... tendered in a tone of voice suggesting that an explanation is clearly needed to allay my incredulousness at such an apparently strange measure. The reaction [would be] fascinating and illuminating. (Adams 1976, 264)

Indeed, in most cases where couples decide to get married, it is more likely that they will be asked *when* and *where,* rather than *why.* This highly ritualized pattern of social interaction yields almost automatic enthusiastic blessings. Asking a couple why they have decided to marry would appear illogical and inappropriate, and would clearly violate conventional behavioral norms and standards of politeness. This predictable encounter includes a well-established social scenario, normative expectations and etiquette norms. Raising this kind of question, as Adams suggests, is considered to be bad manners and a transgression of etiquette rules.

The ongoing disbelief stressing the immense power of this cultural convention continues to mark single, thirty-something women as deficient and incompetent, as well as an object for interrogation, negative evaluation, and suspicion. Despite the prevalence of political correctness codes, the question appears to remain as socially legitimate as ever.

However, as daily interactions attest, such questions can be asked at any time. For years now, I have been asked again and again to justify my status as single and childless. It could happen anywhere: at a friend's wedding, in a cab on the way to the airport, or whilst awaiting my turn at the hairdresser's. In one particularly comic encounter, whilst speaking with three Italian sisters in a Venetian salon, all three raised their hands and shook their heads in disbelief: "Ma tu sei bella, tu sei bella!" (But you are so pretty!). I was thirty-nine- years- old at the time. As "late singlehood" is still perceived as a temporary and "unnatural" category, no efficient silencing system has thus far been developed which would de-legitimize such questions and no etiquette exists concerning what is appropriate to ask. The constant need to account for and justify one's singlehood and agency to control one's timeline is not limited to private social interactions.

One example that stands out especially took place a few years ago, during a press conference in Jerusalem with the American Secretary of State, Condoleezza Rice (Gearan 2007). During the event, Rice was asked about her status as a single woman. Specifically, she was asked whether being single might hinder her capacity to relate to the pain experienced by American families who had lost their loved ones in Iraq. Rice responded, somewhat bashfully, by observing that being single did not render her incapable of understanding American sacrifices in times of war. Following on, Tzipi Livni, Israeli Minister of Foreign Affairs at that time, declared (after first clarifying that she was married with children) that during informal conversations with Secretary Rice, the latter consistently expressed her deep sorrow over American losses in Iraq, and that the American public should know this.

This was not the first time Rice had been obliged to account for her single status. *USA Today*'s reporting on the incident also revealed that at a previous function in the US Senate, Senator Barbara Boxer told Rice that "without an immediate family [she] will pay no personal price for the Bush administration's policy in Iraq" (ibid.). Later, Rice admitted that she was at first perplexed by the exchange, telling Fox News: "Gee, I thought single women had come further than that" (ibid.). These comments, particularity when pronounced by a progressive feminist politician such as Boxer, could be seen as another indicator of the extent to which critical thinking and feminist critique are ignorant of critical studies of singlehood.

In the Introduction, I referred to Garland-Thomson's work (2002, 2), which asserts that disability is still not an icon on many critical desktops: by paraphrasing Garland-Thomson's observation, I have made a similar assertion about the relationship between singlehood and feminist theory and practice. Indeed, feminists have paid scant attention to the ways in which singlism constitutes a form of inequality, reflecting explicit and implicit forms of oppression. The incident with Rice, in common with so many other encounters unfolded throughout this book, reflects how singlism permeates our daily lives but remains rarely acknowledged as such, even in progressive circles. For instance, I believe that it would be far less appropriate, today, to inquire as to the reasons for a woman's sexual preferences on live television in Israel and many European and Anglo-American cultures.

These discursive patterns are often highlighted to an extreme in popular television shows, such as the Israeli versions of *The Bachelor* and *Dating in the Dark* from the UK, for example. Reynolds (2008) describes how in the television program *Holly and Fearne Go Dating*, broadcast in the UK, the two hosts approach passers-by and ask them "Why [they] think [they] are single?" (ibid., 123). As Reynolds clarifies, the show was based on the premise of scrutinizing the relationship history of the single person interviewed, while attempting to discern possible obstacles to the attainment of couplehood and family life. The context created by this reality show presents being single as a problematic state, one that can only be accounted for through personal mistakes and inappropriate conduct (ibid.).

These occurrences highlight the fact that it is still quite socially acceptable to treat singlehood as a legitimate target for suspicion, mockery, or even public humiliation. The press conference in Jerusalem revealed a fascinating cultural dynamic highly relevant to our analysis. What particularly captured my attention while watching this scene was the fact that singlehood, even when attached to such a public figure, remains a category that generates suspicion and demands explanation. Having a family, it appears, still provides many societies with important social signals about a person's constitution and moral character. In common social imagery, this status confers responsibility and credibility. No matter how ridiculous or flawed these assumptions may appear to be, they reflect deeply ingrained social understandings about sociality, collective timetables, and gendered respectability. Indeed, they are grounded in a long legacy of thinking, through which single women have been subjected to disproportionate scrutiny, exclusionary measures, and prejudicial beliefs regarding their character and civility.

These kinds of inquiries reflect, among other things, the social confusion that occurs when singlehood ceases to be a temporary stage. As noted before, singlehood is discursively framed as a liminal, temporary state; a transitory stage on the way to couplehood and family life. In his study on the ritual process, Victor Turner (1969), drawing on Arnold Van Gennep's (1960) theory of the three stages of rites of passage, placed particular emphasis on the second stage—the liminal stage, highlighting its fundamental social importance. Liminality, he emphasized, is a state of being between phases—a transitory state. The individual positioned in the liminal phase is neither a member of the group she previously belonged to, nor a member of the group she will belong to upon the completion of the next rite.

In similar fashion, lifelong singlehood marks an unexpected disruption, a normatively liminal state which has unexpectedly become a permanent one. This stage, as Turner (1969) notes, is also characterized by ambiguity and inversion resulting from an anomaly wherein people slip through networks of classifications. More specifically, singlehood at the age of eighteen or twenty-three is still located within the boundaries of the socially structured life course, whereas it would fall outside of this framework after a few short years.

I consider the ever repetitive question "Why is she still single?" as reflecting some of the dominant social temporal understandings of singlehood. At some point in the single woman's life trajectory, her singlehood shifts from being a socially legitimate temporary phase to what can be considered as a biographical and social disruption (Bury 1982). This leads us to some of the central questions in the present study: What are the discursive implications when singlehood ceases to be temporary and becomes a permanent status? Why is it almost impossible to imagine one's timeline beyond hegemonic heteronormative norms of reproductive linearity? Which discursive templates structure and cultivate these barriers? Likewise, why is it so difficult to conceptualize singlehood not merely as a liminal phase but as long-term life option? What happens when the expected passage to the next stage in a life-course structure is delayed or does not occur at all? And which discursive forces bestow these temporal truth claims with so much power, privilege, and universal normativity?

I view the question "why is she single?" then as a response to a transgression of socio-temporal boundaries reflecting the social confusion that occurs when singlehood ceases to be a temporary stage. One of the evident ramifications of this transgression is the need to provide explanations and justifications for this unexpected disruption. Interestingly, marriage, couplehood, and parenthood, according to common-sense knowledge, still serve as indicators of coherence, meaning, and moral order. These widely held beliefs also construe a particular kind of subjectivity.

Beyond the prevalent stereotypical labels attached to single women, they are depicted as leading empty, meaningless lives, and as lacking in moral competence and character. Not only are single women subject to increasing suspicion concerning their dubious moral trait, and not only do they "fail" to account satisfactorily for their singlehood, but their singlehood positions them as accountable to no one, and no one as accountable for them. This could also possibly explain why Boxer's critique of Rice did not trigger a media storm. I will continue to explore these threads in the next chapter by offering a critical reading of age, ageism, and singlism, as a potential source of invaluable insights to some of the taken-for-granted assumptions concerning temporality and singlehood.

Notes

1   For a rich analysis of the popularity of bachelor and bachelorette parties see Montemurro (2003, 2006); Tye and Powers (1998).

2   I draw on Goffman to stress, as he does, the "moral aspects of career – that is, the regular sequence of changes that career entails in the person's self and in his framework of imagery for judging himself and others" (Goffman 1972, 258).

3   The pseudonym of two single women columnists writing for *Ynet*.

4

Facing the horror: becoming an "old maid"[1]

The blatant contradiction that exists between the terms "old maid" and "young single woman" is not merely anecdotal data from the flippant lingo of contemporary popular culture, but rather a significant cue for understanding the tenor of our times. Despite dramatic changes in family lifestyles coupled with growing numbers of single women, the well-worn myth of the aging single woman as a miserable yet terrifying old maid appears to have resisted these trends. Rather, the myth persists, as a naturalized, undisputed, and insoluble cultural trope. Indeed, cartoons, jokes, and horror stories about "old" single women are widely accepted and disseminated, a cautionary reminder for women concerning the specter of being single in old age and what looms ahead for them. In that light, single women are often the subject of caustic remarks, sardonic humor, patronage, and scorn, because they are seen to pose the constant threat of pervasive perversion to the normative societal order.

This chapter asks what gives this powerful stereotypical image so much discursive force and makes it so defiant to resistance and deconstruction? Addressing recent literature on age, feminist theory, and singlehood, I investigate the ways in which ageist and sexist constructions of age form prevalent understandings of lifelong singlehood. It is my contention that single women above a certain age are faced with a triple discrimination, based on their age, gender, and single status. In this chapter I examine the manner in which the language of age guides common-sense understanding about single women. Specifically, I explore how the predominant cultural perceptions of age appropriateness, age segregation, age norms, and ageism play a crucial role in the construction of lifelong singlehood and gendered timetables in general. Ageism and age-based discrimination, I argue, do not necessarily apply merely to the social category of old age, but are practiced at different stages of the life course.

A new analytical perspective will allow for prevalent perceptions on age identity, age norms, and age relations to be placed in context. This chapter considers, in particular, where identity vectors like age, gender, and relationship status converge, and notes that questions such as "Why are twenty- and thirty-plus single women depicted as old?" and "Why are thirty-plus married mothers represented as 'young mothers'?"

emphasize that single women are aged by societal norms determined by culturally framed expectations.

In what follows, I attempt to unpick the discursive process which causes single women to "age faster": why do single women age differently from coupled and married ones? Indeed, ageist assumptions also tend to prevail in popular discourses about late singlehood and the categorization of the "aging single woman." This is why this chapter proposes an analysis of the aging process of single women as a socially situated symbolic practice and not—as it is customarily grasped—as a given biological category.

Age and singlehood

In her analysis of single women in popular culture, Anthea Taylor (2012) proposes that the study of single women opens a window on how heteronormative and patriarchal frameworks operate in new and sophisticated ways. Inspired by Taylor's study, I contend that current categorizations of the "old maid" are deeply embedded within the context of heteronormative culture. According to Berlant and Warner, heteronormativity is:

> The institutions, structures of understanding and practical orientations that make heterosexuality seem not only coherent—that is, organized as sexuality—but also privileged. Its coherence is always provisional, and its privilege can take several (sometimes contradictory) forms: unmarked, as the basic idiom of the personal and the social; or marked as a natural state; or projected as an ideal or moral accomplishment. It consists less of norms that could be summarized as a body of doctrine than of a sense of rightness produced in contradictory manifestations—often unconscious, immanent to practice or to institutions. (Berlant and Warner 1998, 548)

The privileging of heterosexual and familial bonds has the pro-active force of structuring normative understandings about single women and aging. In this manner, as Taylor (2012) writes, single women are situated in relation to, and as against, the married/single binary, and are construed as figures of profound disparity. These sets of assumptions become ever more unforgiving as single women age. In their discussion of single women's accounts of their single status, Anna Sandfield and Carol Percy (2003) note how references to older single women are generally derogatory, as well as how older single women are perceived as lonely and isolated. According to Sandfield and Percy, all the participants in their study demonstrated an awareness of the status-related expectations associated with age (ibid., 480). I concur with Sandfield and Percy's findings, and stress that socially produced consciousness is embedded in the age conventions guiding mundane social interactions, and plays a key role in the discursive construction of thirty-plus single women.

Scholars such as Hazan contend that the omnipotence of age is revealed in the fact that age is perceived to be an objective, universal, natural fact, and beyond dispute:

> Age is mistakenly considered to be a universal category. Although it is often endowed with the analytical status of a "variable," it appears as something which could not be

explained ... This mistake stems from a lack of critical deconstructive thinking about the concept of age. The identification between biological, social, psychological and chronological age is affirmed in developmental psychological theories which constitute age clusters at different stages of the life course and bestow age with features which are beyond its classificatory marker. (Hazan 2006, 82)

In his writings on the reasoning of bureaucratic logic, Don Handelman discusses the effectiveness of age as a taxonomizer which constitutes the temporality of the individual, "smoothing him into the bureaucratic order" (Handelman 2004, 88). Accordingly, each life phase defines its own age-appropriate behaviors, and serves as a key tool for producing knowledge, coherence, and meaning. As Handelman suggests, "Knowing one's own numerical age—one's exact location in time, synchronized precisely to all other individuals—is considered an elementary index of competence" (ibid., 59).

Prevalent images of single women suggest that passing, or being around the age of thirty demarcates a crossover zone. In this sense, the knowledge of one's age discursively constitutes the single woman's status, and provides allegedly significant evidence for determining who the single woman is and what she ought to be. This also stands in tandem with Cheryl Laz's (1998) research on the performative aspects of age. Laz views the category of age as an accomplished one, or as she puts it: "We collectively do it right" (ibid., 99). I can locate my book within the broader visions of feminist theorizing about aging. A common-sense view embedded within our patriarchal and youth-oriented culture is that as women age, they move away from current beauty ideals, and accordingly need to develop age concealment techniques.

As Catherine Silver observes:

> Older women's bodies are more likely to be perceived as deformed, ridiculous looking, and desexualized. They become frightening, "crones" and "witch like," as imagined in children's books and fairy tales. The language that describes older women is indicative of deep-seated, unconscious fears and a rejection of the ageing female body, with its connotations of danger and contamination that need to be kept separate and isolated. (Silver 2003, 385)

Silver's reflections accord with Susan Sontag's statement in her celebrated essay, "The Double Standard of Aging":

> [Women are considered] Maximally eligible in early youth, after which their sexual value drops steadily; even young women feel themselves in a desperate race against the calendar. They are old as soon as they are no longer very young. (Sontag 1983, 102)

Here, Sontag gives us insight into the deeply ingrained symbolic order that defines a single woman from a certain age as "no longer very young." Sontag's explanation is also especially relevant to understanding the gendered aspects linking aging and singlehood. The single woman's aging process is a marker of her gradual withdrawal from the market, signifying her diminished sexual and reproductive value and functions.

Debating the thirty-plus-year-old "old maid"

In what follows, I seek to understand some of the discursive mechanisms by which the pejorative "old maid" label continues to be reproduced. The following analysis shows that the well-worn trope continues to prevail, in contemporary Israeli culture as well as in many societies where singlism reigns supreme.

Orit Gal, a single woman writing on the *Ynet* portal claims:

> From a certain stage, every single woman will be tagged as a shriveled old lady. She will be pitied by her surroundings including her friends, family and colleagues for being an old hag. She will pass her nights by watching television, eat without control and share her bed with cats as no normal men would want to touch her. (Gal 2010)

Gal refers to the transition point through which women turn into old maids. The very process by which single women "age faster" than their married counterparts is loaded with sexist and ageist assumptions. Deeply entrenched within these presumptions is the perception of the single woman as a site of danger and contamination:

> [The image of an "old maid"] is a warning signal that embodies the cruel destiny which awaits a woman who remains single. She might find herself cast as the "crazy cat lady"; this aging, solitary, poor woman who hangs around the neighborhood with her night gown on and feeds all the neighborhood cats. (Banosh 2011a)

As Noa Banosh, the single woman whose column was published on *Ynet*, comments, the image of the single woman as the "crazy cat lady" is one of the more common stereotypes that crosses cultures and time. Therefore, it is not surprising that columns and commentary like the above turn to this specific image when predicting the future awaiting single women. The common reference to cats is worth mentioning; indeed, as single scholars like DePaulo (2006) have observed, the unmarried woman is regularly stereotyped as lonely, miserable, and with no alternative but to fill her empty life with cats. Thus, the presence of cats have come to symbolize the lack of men in single women's lives, as by this point in their lives they only have cats to keep them company. Moreover, this could be seen as a metaphorical representation of the inferior status bestowed upon single women by society at large. As feminist scholars have argued, this association of women and animals resides within a patriarchal, heteronormative conceptual framework, one which justifies the domination of women and the superiority of men over them, as they are presumed to be more primal and animalistic than men (Donovan, 1995; Spelman, 1982).

These accounts provide insight into the ways single women internalize widely held views about single women. Although often executed with humor and irony, by referencing this set of images, many single women embrace the typical image of the aging spinster living alone with her cats; to a certain extent, they even participate in keeping this image alive. By doing so, they also observe themselves through a patriarchal and sexist gaze, through which they become dominated and objectified. Hence, even though they realize that this image functions as a disciplinary mechanism, they cannot resist the cultural scripts which refer to long-term singlehood in terms of emptiness, loneliness, and loss. This formulation conveys a horrendous future: if they don't find a partner at the right marriageable age, they will end up living a lonely, mentally unstable, and socially marginal life.

The image of the crazy cat lady also represents the pathologization of older women in our society, women whom, as Silver (2003) notes, should be isolated because of the fear of contamination. The above quotations also exemplify the process through which women internalize the normative gaze to which they are subjected. This is reminiscent of Sandra Bartky's explanation concerning how women subject themselves to the normative gaze and judgment of men:

> In contemporary patriarchal culture, a panoptical male connoisseur resides within the consciousness of most women: they stand perpetually before his gaze and under his judgment. This is a process through which they become isolated and self-policing subjects which internalize the male normative gaze and are controlled by it. (Bartky 1990, 72)

Through an adoption of the male connoisseur's panoptical gaze, the loss of youth, beauty, and reproductive power turns women into social rejects. Within the context of this study, single women can be expected to experience relentless anxieties about their age, beauty, and reproductive abilities. Clearly, single women above a certain age cannot possibly compete with younger women, given that they are on the verge of losing what are considered women's most important social assets: their appearance and their reproductive potential. This particular form of age hierarchy will be further explored through what I describe as the single woman's accelerated aging process.

Accelerated aging

Central to our discussion is the manner in which sexist and ageist beliefs produce a particular kind of accelerated aging. The data analysis indicates that to a certain extent, single women "age faster" than married ones, and it is this very symbolic social process that contributes to the stigmatization and devaluation of single women. This analytical concept demonstrates how we are aged by culture and narratives about time (Gullette 2004), and sheds light on how perceptions of the aging process are determined by age-appropriate behavior and age norms. My use of the term "accelerated aging" draws from a study about aging among gay males, conducted by Keith Bennett and Norman Thompson (1991). In their study, they argue that:

> Homosexual men are considered middle-aged and elderly by other homosexual men at an earlier age than heterosexual men in the general community. Since these age-status norms occur earlier in the gay sub-culture, the homosexual man thinks of himself as middle-aged and old before his heterosexual counterpart does. (ibid., 66)

In a similar vein, Julie Jones and Steve Pugh (2005) contend that in a society where ageism and homophobia are endemic, to be old is bad enough; but to be old and gay is to double the misery. Jones and Pugh's observations can be extended to the study of single women: *to age as a single woman triples this misery*. Bennett and Thompson's research joins other studies that have analyzed different forms of premature aging, such as with ballet dancers, table dancers, and athletes (Ronai 2000; Turner and Wainwright

2003). For example, Carol Ronai's (2000) study of "aging table dancers" examines the social process through which table dancers are perceived as being older at a relatively young age.

> As chronologically young as she [the dancer] may be, she can be old. Her body is not as supple and her dance not as animated as it once was. Her gestures toward customers are construed to be abrupt, demanding, nagging, less patient than before. A dancer's sexual utility and the sincerity of her presentation come into question. (ibid., 315)

Drawing on these observations, I found that the concept of the thirty-five-year-old single woman is a vivid expression of accelerated aging, which in turn construes different timetables and rhythms for single women. This process is vividly exemplified in the next column, written by Tal Hashachar, a single woman and columnist on the *Ynet* portal:

> You are already twenty-three years old, you better not rest on your laurels—beauty does not last. You should begin to compromise ... You better understand honey, that women age and men grow up. Very soon you will be considered an old maid and you ought to begin to think about a name for your cat ... If you don't compromise, and as soon as possible, it will be catastrophic. And if you're not married or on the path to marriage by the age of twenty-five in a magic spell you will realize that you have turned into an old and ugly maid and feel remorse about all the ugly ducks that you have rejected in the past whom by now have turned into swans without you. (Hashachar 2011)

This account illuminates how the process of accelerated aging takes place even when one is twenty-three years old. To some extent, the author echoes the feminist critique on age and aging when she states that "women age, men grow up." She is very much aware of the gendered process of aging and the hierarchical relations that this produces. In this context, her analysis is reminiscent of Susan Sontag's (1983) and Judith Gardiner's (2002) observations, that the aging processes of men and women are culturally marked in highly asymmetrical ways. This process of devaluation is based upon the premise that a woman's value is dependent on her appearance and reproductive capabilities. Hence, the warning addressed to single women is clear: they cannot rest on their laurels, as evidently they are in danger of losing their ability to perform as objects of sexual desire and to fulfill the role of future mothers.

For that reason, single women are obliged to compromise.[2] As Tal explains above, the men whom single women rejected in the past have now "turned into swans": that is, as the "market value" of a single woman decreases, that of a single man increases. According to this perspective, men age well and younger single women possess a natural superiority merely by virtue of their age and gender status. Giga, another columnist writing on the *Ynet* portal, also reflects upon the stigmas attached to her age and single position:

> I am thirty-six; the truth is that I'm almost thirty-seven. So come on, you enlightened men; hang me at the outskirts of the city and don't forget to hang above me a sign denoting that I am an old single woman, a "rotten tomato," "damaged goods," or something similar. I am sure you have an abundance of nicknames for girls my age. And to the

women who have not experienced the dubious pleasure of being single above the age of thirty-five: continue nagging me with fertility tests and stories of single motherhood, menopause, and the state of my ovaries. This will definitely help me find a groom tomorrow. (Giga 2007)

Being called a "rotten tomato" or "damaged goods" alludes to the age-based market from which single women are in danger of exclusion. This perception of "the single woman's short shelf-life" will be further developed in the next chapter. However, within this context I wish to stress how the cult of youth is given absolute priority. Accordingly, women are socialized from early stages in their lives, to be wary of losing their beauty, sexual desirability, and reproductive functions. Such a loss will most likely disqualify them from competing in the heteronormative dating market. The temporal logic of the market is articulated as an absolute, timeless truth, one which abolishes all other social experiences. These claims are cast as deterministic, whereby single women have to adjust to laws of supply and demand. A single woman within or beyond the marriageable phase should be particularly cautious about her aging process.

In this light, the continued presence and threat that the "old maid" represents in the public imagination reaffirms the heteronormative, familial, and age-obsessed ethos. The fear which this image evokes can be linked with what Sherryl Vint (2007) describes as a new kind of backlash, one which frightens women into accepting traditional gender roles and convinces them that their lives should be focused around heterosexual marriage and motherhood. In fact, the construction of the old maid as a source of collective fears bestows more ideological force to the idealization of the conjugal and maternal bonds, and construes neo-traditional models of the post-nuclear family.

Complying with the heteronormative and familial models represents successful timing. This perhaps can shed some light upon why thirty-plus mothers are called "young mothers," while single and childless women of the same age are termed "old." The label of "the crazy old hag" or "aging old maid" is another indicator of how the chronological aging process of women is embedded within heteronormative, ageist, and sexist assumptions, through which they are devalued and socially marginalized. In this context, this stereotype designates the social death which awaits them, a conceptualization which will now be further developed.

Social death as a solution to the insoluble

David Sudnow (1967) described social death as a prelude to biological death, which usually begins when the physician gives up all hope of a patient's recovery, and puts a time limit on the patient's survival prospects. At this point, the institution loses its concern for the dying individual as a human being, treating her as if she were already dead.

Several of the single women columnists on *Ynet* identify themselves as approaching what may be considered as their own social death. In this case, social death denotes the single women's diminishing market value, through which they cease to be worthy sexual subjects and in turn gradually lose their value in the dating marketplace. In this context, Merav Resnik, a regular writer on the *Ynet* portal, explores her experience of visiting Israeli and American internet dating sites:

> Beauty is irrelevant; the world belongs to the young. Whoever told me that I'm still young was a big liar. On this American dating site there is an age limit, but as opposed to the age limit we are familiar with from clubs and pubs, here the upper age limit is the one that counts. Anyone above thirty-five does not exist. I have two more years to "live." It does not interest them that they could be missing out on someone who can be beautiful. If she's "old," she's out. (Resnik 2007b)

The aging process, as Merav asserts in the above quote, is reminiscent of social death: the "upper age limit is the one that counts" and according to these age norms she only has two more years to "live." This age-stratified boundary indicates the writer's risk of future exclusion from the marriage market. In addition, these rules of supply and demand are beyond her control; as Merav declared, "If she's 'old', she's out." According to this socially determined timetable, the thirty-plus single woman is "left on the shelf," and ceases to live; she is out of the game. According to the logic of this particular market, single women must be careful managers of their time, and this objective can only be achieved by abiding to patriarchal, ageist traditional discourses. In this undisputed economic rhetoric, the categories of age, gender, and one's relationship status trivialize and denigrate the lives of single women.

These discourses are closely connected within the general perception of old women. As Diane Garner puts it, women lose their social value simply by growing old (Garner 1999, 4). This contention is further developed by Silver:

> The female body, which no longer reflects reproductive abilities nor attracts "the gaze" of men, has become a reminder of death to come. The fears of ageing and death have to be controlled and kept at bay, especially in a society like the United States, which is obsessed with youth images, narcissistic gratifications, and the prolongation of life at all costs. (Silver 2003, 386)

The next writer, Lotti Kremba (a pseudonym), a thirty-nine-year-old single woman and columnist on the *Ynet* portal, unfolds this process:

> I am a thirty-nine-year-old woman. I live in an allegedly Western modern society in the new Middle East. I live in Tel Aviv and not in a remote religious village. Yet, I'm surrounded by a substantive number of people who believe that being almost forty years old without a husband and children is tantamount to having one foot in the grave. Am I being too extreme and dramatic? Am I overreacting? Not at all, in fact, on dating websites men stopped acknowledging my existence when I turned thirty-five. It's hard to see the people behind the number … In the real world things are not much better. Men treat me nicely only to take me to bed, as you know what they say about "experienced women" my age … According to common knowledge, older women are great in bed … in this and other senses, the talkbacks mock single women who have passed the age of thirty and have remained single. (Kremba 2009)

This account corresponds with parallel experiences of older women, whose invisibility is symbolic of their social exclusion and isolation (Woodward 2006). Indeed, the two accounts above reveal how the writers experience their singlehood as a form of social death, through which reaching this age without a husband and children is "tantamount to having one foot in the grave." As Merav Resnik (2007b), the previously quoted columnist, declared, "anyone above thirty-five does not exist," given that she has only "two more years to live."

The experiences of invisibility described by these single women also resonate with Goffman's (1959) definition of "non-persons." Goffman viewed non-persons to be "a standard category of people that are sometimes treated in social interactions as if they are not there" (ibid., 152). According to these narratives, thirty-plus single women gradually become such persons. Although they wish to participate in the dating/marriage market, they are ignored and rendered invisible.

The writers place much emphasis on their age, and are highly aware of the manner in which they are evaluated and objectified in accordance with their age. Both writers referred to being thirty-plus or approaching the age of forty as a crucial age marker separating the visible from the invisible, social life from social death. Their age becomes their *indivisible master status* (Becker 2008) or their *master determining trait* (Hughes 1971), which tends to overpower the other characteristics which run counter to their biological age. In this manner, they defy the prevailing age norms and societal expectations through which a woman's role is pre-determined: marrying and having children at the right age.

Age-appropriate expectancies

Occupying the position of a single thirty-plus-year-old marks a transgression of age-appropriate behaviors and expectations. It appears that single women are classified and stratified by their age rank, which places both the twenty-something single woman and the thirty-plus married one higher on the normative social scale. These cultural assumptions corroborate Marlis Buchman's (1989) assertion that age creates different social categories, defining qualities, rights, obligations, and motives associated with members of a given age group and forming hierarchal relations between them.

The transgression of age-appropriate expectations could be traced in the next passage, written by Inbal Bli K'chal Vesarak, a single woman and a columnist writing on the *Ynet* portal:

> I have decided that until I have a steady partner to show up with to Friday dinners, I'm not going near my family's house. Although they don't ask, I can see the question marks flickering in their eyes: "Well? When? You are almost thirty-seven!" (Bli K'chal Vesrak 2008)

The age thirty-seven, in this case, is constituted as a symbolic checkpoint introducing new tensions between the writer and her family. Inbal writes that being a single woman who is almost thirty-seven requires a "special travel permit"—in this case a date, a boyfriend who may signal a potential promise of a husband and father-to-be. The above passage is another illustrative example of how age norms are entrenched in everyday life, forming rigidly age-scripted social expectations and interactions. The required injunction to successfully adjust to these age norms can also be found in the next extract:

> Oh how much I feared this age, thirty years old. I stopped breathing every time I thought about it. So far I'm thirty years old and one month. With hesitation, I can say that this is not such a bad experience. Yes, I reached the age of thirty as a single woman and I'm still alive. The only marks of scorn arise from my family, who have already begun to question my sexual preferences. I hear nasty remarks such as "Even your cousin found someone, why can't you?," "Nothing is ever good enough for you," "Why are you not ready to be fixed up?" I am ready, I'm definitely ready but why should I date someone who is fifteen years older than me? The fear from my age and staying alone freezes me and often makes me lose my capacities for logical reasoning. [These fears] also make me forget who I am, to stand up on my own and be respected. (Or 2011)

Most of the writers understand the significance of their changing age status, and are aware of its attendant social ramifications. The extracts above highlight the manner in which one's age identity, after a certain point, becomes a potent mode of regulation, a disciplinary mechanism which evaluates the single woman's behavior, personal attributes, and social worth. Thus, what is often a socially acceptable status in one's early twenties rapidly transforms in into a category that is subjected to increased social scrutiny. Rotem Lior, a single woman writing for *Ynet*, is also aware of this transition and its ramifications:

> It seems that men [who are thirty-plus] do not consider in any way women of that age as eligible, calling them instead "flattering" names such as "bitter" and "uptight" … What men run away from, according to their own accounts, is the desperation that those women express and not their age or biological clock. This very desperation is what makes them run—screaming—straight to the young breasts of twenty-year-olds. (Lior 2007)

As these extracts demonstrate, the value of single women is determined by the evaluative gaze of men, a gaze which objectifies them according to their age. Moreover, the ageist and sexist market rules of supply and demand produce hierarchical age relations. The cultural preference for younger single women is also set in commonplace preconceptions about the preferable age gap between men and women, and reflects a sexist and ageist gendered social order.

Another example can be found in the next column, written by Sivan Stromza and published on the *Saloona* portal:

> My name is Sivan Stroumsa, I am a thirty-year-old single woman, almost thirty-one. Why am I single? Because. That's where my life has led me to, fate, circumstances etc. No, I have nothing "wrong" with me. Everything functions properly (apart from my BlackBerry). No, I'm really not a model, I am five feet and have small breasts, but I dare I say I am quite a catch, perhaps even much more than many women who are younger than me, the "normal" kind, who married at the "right" age or at least have a partner. (Stromza 2012)

Singlehood is conceptualized through the polarized terms of young versus old, women versus men. In the column quoted above, Sivan is fully aware of the privileged category of normative youth. The accounts above also confirm another social convention often taken for granted: it is acceptable for older men to date and marry younger women, but not the other way around. This "common knowledge" reflects how age and gender are fundamental organizing principles which still place substantial limits upon the life trajectories and options of single women. It also explains why so many fears, anxieties, and social pressures are endemic to the aging process of single women.

Age, then, is perceived as an essential trait, an absolute status that defines its own expectations and capabilities. As such, each life phase designates its own age-appropriate behaviors, and serves as a key tool for producing knowledge and stereotypical labels. Age appropriateness norms serve as a crucial parameter for constituting one's persona and life trajectory. The single woman is depicted as living outside the normative life cycle. Instead of moving ahead in a linear and a sequential fashion, she is moving "over the hill." Late singlehood symbolically draws single women closer to the life stage of old age and death. In some respects, they metaphorically skip "mid-life." Their singleness ages them exponentially and prematurely. In many ways, according to traditional discourses on the single woman's life-course from now on, their life span is endowed with clarity and certainty: aging alone and dying alone.

The tyranny of age

The data analysis thus far demonstrates the power of familial, heteronormative, ageist, and sexist norms in construing the category of the "aging single woman." However, some of the single women writing for *Ynet* do not conform to these hegemonic discourses, instead challenging them in various ways. One apt example is a text written by Rotem, one of the columnists mentioned above, who defines herself as "a thirty-something content single woman" (Lior 2006). In her column, she describes what she terms as the hysterical behavior of her friend Maya, a thirty-plus single woman who is eagerly looking for a husband:

> My friend Maya is a single woman, a thirty-plus that has turned into a minus. People raise their eyebrows and she plucks them ... Maya is a member of the "groom sect." At this rate she will become a nonprofit organization and start looking for funding. I reminded her that I'm also a thirty-plus single woman but I'm pretty much enjoying it ... she lost her temper and threatened to hospitalize both of us in a mental institution; she would be hospitalized for her singleness and I would be hospitalized for my unstable mental state. (ibid.)

Attesting to the powerful force of binding age norms, Rotem first situates herself as a thirty-something content single woman. By this, she acknowledges the stigma of the thirty-something miserable single woman, and aims to defiantly subvert this age-based symbolic order. Furthermore, the stigma is transformed in her case into a positive form of self-identification and alters the controversy surrounding that very symbol. Rotem's statements pose an alternative to the prevailing image of the thirty-plus "cat lady"; she makes it clear that she enjoys her status. By this very statement, she unsettles the basis of hegemonic heteronormativity, which assumes that the joy and meaning of life can only be found in getting married.

In another column, Hadas Friedman, also a single woman and a writer on *Ynet*, makes a similar point:

> I carry with pride the title of a thirty-five-year-old single woman who lives in Tel Aviv (but with no cats, as my dog won't allow them). I must say that it's pretty nice to be a thirty-five-year-old single. What is less pleasant are the stereotypes about the aging single woman. If you meet three different thirty-five-year-old single women and ask them to tell you about their romantic life stories, every story would be different ... You would meet very different women with regard to look, character, personal taste in men, and plans for the long and short-term. Not all of them are obsessed with marriage and having children. (Friedman 2009)

Both columnists challenge some of the dominant beliefs regarding the thirty-plus single woman. Hadas stresses that she is proud both of her age and single status, and in this way resists dominant images of aging, femininity and singlehood. However, she is aware—as are most single women—of the stereotypes that single women are subjected to. Within this context, she notes that singlehood is more diverse than its stigmatized images so deeply rooted in the attitudes of her surrounding social environment. According to Hadas, if one would simply bother to ask single women about their lives, one would discover narratives which do not necessarily adhere to the cultural script of the aging and miserable single old maid. Here, she pinpoints how single women find themselves subject to reductive and essentialist representations, and underscores how the lives of single women are much more diverse than the prevailing representations.

The next account, taken from a different column, also attempts to challenge the stereotypes and pathologization of single women as they age:

> Recently, I read an article about the different remedies that are supposed to cure the single disease for people above the age of thirty-five. While reading the talkbacks, I couldn't help but notice that one of them exclaimed: "Anyone that is single above the age of thirty is damaged goods." I smiled to myself; how lucky I am to be only thirty-two. My damage is considered to be light; I am safe for the time being. But the thought hasn't really disappeared. (Resnik 2007c)

In fact, the reading of the texts reveals how the struggle against common-sense assumptions concerning thirty-plus single women is often expressed through a complex interplay of power and resistance, compliance, and confrontation. Merav relates to how her age and single status are perceived as a disease, an individual and social pathology. Like many single women writing on the *Ynet* portal, she relates to these attitudes with an ironic tone, which allows for a self-reflexive critical distance. That is, her humorous readings, like most of the other texts quoted here, also reflect a relational and situated understanding of how singlehood is constructed according to idealized and rigid types of femininity.

At one level these texts can be read as acts of resistance, but at the same time they also accept and endorse hegemonic definitions of age, gender, and the required course of life. The ironic tone also assists us in recognizing the polysemic nature of the texts, in which the tenets of an ageist, heteronormative culture are both resisted and endorsed.

Most of the accounts analyzed in this chapter criticize the stereotypes associated with the category of the old maid. Yet, it is important to note that the objective, natural, and scientific qualities of age, as well as the downward mobility associated with aging, are reluctantly acknowledged. As Merav Resnik discloses in the above account, the troubling thought about her aging as a single woman refuses to disappear. Merav's experience reflects how these beliefs are internalized by those who are stigmatized, leading to stress and anxiety. I would like to propose that the ways in which these skeptical stances are generated and divulged demonstrate the discursive power of age, and the powerful authority it bestows, as well as the pejorative qualities assigned to midlife (Gullette 1998) and old age (Hazan 1994). This concurs with Hazan's (2006) contention that the disciplinary power of age scarcely provokes any criticism and protest. The discussion about single women stresses the need to transform existing conceptions about age, aging, and feminine subjectivities. Hence, in order to pave a way for forming a new counter–discourse about women's expected life trajectory, more scholarly attention should be given to questioning the perception of fixed and stable age identities, and how these are culturally constituted and reproduced.

Both popular and scholarly literature often refers to "the thirty-plus single women" as a collective and worrisome phenomenon. Single women are categorized according to their age cohort, and are denoted accordingly with essentialist attributes and collective behavioral patterns. In this chapter, I have attempted to contribute to a more nuanced understanding of the discursive construction of the "aging single woman" and singlehood in general. My intention has been to dispel current pervasive opinions, and to call for more scholarly research into how shared beliefs on female singlehood and aging are endorsed and followed.

In contemporary discourses on single women, age appears to be the crucial reference point—albeit one often taken for granted uncritically—with which women are measured and evaluated. Age categories crystallize biological and social qualities into unquestioned roles and norms. The discursive analysis of singlehood in this study demonstrates how far we are from living in what Michael Young and Tom Schuller (1991) term as an ageless society. The symbolic language of age and family life as an archetypal collectivity carries with it a system of meanings through which women are defined, express themselves, and interact with others. Therefore, rethinking the discursive formations of the category of the old maid, family, aging, and age reveals prevailing conceptions of an essential and normalizing social order, which rest upon allegedly objective and valid inferences. From this perspective, couplehood and family life represent the promise that order, coherence, and meaning will be bestowed upon one's life trajectory, while singlehood is stripped of such, leading one to social isolation and loneliness.

We can now see the various forms through which age, gender, and marital status categories intersect. Age defines, regulates, and produces knowledge, and occupies a central place in the socially contingent discursive formations of singlehood. Moreover, the present research emphasizes that the social construction of age and ageism are relevant not only to carriers of the cultural tag of old age (Hazan 2006).

This chapter also attests to a still relatively under-researched process through which, as Ronai (2000) observes, aging is separated from old age. In this vein, the ubiquitous presence of age in the discourse concerning singlehood reveals, among other things, a fascinating social course of action which demonstrates how aging and ageism are not only confined to the further end of the life-course spectrum, but also solidify, through regimes of horror, shame, and guilt, the naturalized and insoluble categories of young female adults.

To conclude, one of the key puzzles that emerges from this chapter is the question "What turns age into such a potent social category, so resistant to criticism and decon-struction?" This question is particularly relevant given the growing concern within the academic discipline of Women's Studies with deconstructing gender, race, class, eth-nicity, and sexual orientations as absolute and natural categories. To elaborate, over the last few decades, social critiques have demonstrated how allegedly pure analytical categories guiding social inquiry and the popular imagination are situated and contin-gent. Furthermore, while postmodern social inquiries attempt to critically dissolve existing boundaries of social categories, the prominence of age and personal status reflect some of the boundaries and margins of these very attempts. Both still operate as core constitutive categories of age and gender-bound social and normative orders. Consequently, feminist age and singlehood studies are yet to occupy a central—and much deserved—place in current feminist theory. Thus, an additional question which begs to be asked is: "Why is personal status, as age, so resistant to deconstruction?" Based upon these observations, I suggest that conceptualizing single women as carriers of the cultural tag of singlehood can illuminate more discursive dimensions, and can open up new avenues for social analysis, both for critical feminist age studies and for the feminist study of singlehood and women's lives in general.

Notes

1   This chapter was written together with Haim Hazan.
2   For my discussion of singlehood and the need for compromise, see Lahad (2013).

5

On commodification: from wasted time to damaged goods

A few years ago, Princeton alumna Susan Patton (2013), sparked intense debate when, in an open letter to *The Daily Princetonian* (Princeton's university student journal), she suggested that female students make the best use of their time at the university by finding a future husband. The only good men out there, she explained, were to be found exclusively in their undergraduate classes. In a follow-up interview with the *Daily Mail*, Patton added that college-age women "have to start putting in place plans for their personal happiness, because they will never again have this kind of concentration of extraordinary men to choose from" (Whitelocks 2014). One sentence that warranted particular attention was her assertion that nowadays, career women "are wasting their youth on caring about their jobs" (ibid.). Elsewhere, Patton has been quoted as saying that "a woman looking for a husband in her 30s gives off total desperation, … No matter that the median age for a woman's first marriage in the U.S. is 26.5 years old—once you hit 30, apparently, it's all over" (Bahadur 2013).

Patton's letter generated much media attention. While some commentators criticized her for a sexist and elitist outlook, others saluted her for her courage in telling the truth to single women. Following on from the controversy, Patton (2014)—who has since come to be known as the "Princeton Mom"—published a book, *Marry Smart: Advice for Finding THE ONE*. The message of Patton's book, in common with her letter, is that a woman's most important life goal is to get married and have kids. Elsewhere she summarized her ideology thus: "So you're 35, who are you going to be looking at to marry? I'm going to say most women who are 35 are going to be looking for a man around the same age, or maybe a year or two older. So let's take the man of 36. He's quite happy to actually be with a woman 10 years younger" (Wallace 2014).

In Israel, Patton's message was mentioned in an article about a short film of a 35-year-old single Jewish woman (Domkeh 2014). Claims like Patton's are ubiquitous in mainstream Israeli discourse about midlife single women. They convey deep-seated assumptions about the need to teach single women the "facts of life" and "how the world really works." One striking feature of this discourse is that these numerous instructional and regulative codes are anchored by temporal economical images, metaphors, and principles. Thus, single women are warned not to *waste* the "*best years of their lives*," to *manage their time wisely*, and to *invest* in the right kind of men/commodity. Moreover, this discourse is imbued with "objective" calculations about man-shortage, consumer demands, expiry dates, and the rules of supply and demand.

Patton's letter is a useful opening point for this chapter because it raises some of the important questions that motivate my analysis: how does the commodified language of time shape our perceptions of female singlehood? What sort of exchanges take place between single men and women, and what are their conditions? How far does the abstraction of time into a quantifiable measure control single women's lives? And what are the discursive mechanisms through which single women become "damaged goods"? In this chapter, I argue that the naturalized, objective temporal rules of supply and demand are significant discursive resources in everyday discourses, and through these the oppression of women occurs and age-gender based hierarchies are produced and maintained.

Within this context, I stress the need for alternative ways of thinking about singlehood, in a manner that disconnects singlehood from the normalized concepts of market logic, exchange value, and the notion that a single woman can be "sold" and "traded." Undoubtedly, there is a need to incorporate feminist thinking into this discussion which aims to debunk the way this commodification of time regulates women's consciousness.

Single women in the temporal marketplace

The single consumer marketplace is replete with a rich vocabulary and descriptive metaphors. These evaluations permeate numerous online articles, which depict the marketplace in terms of a demographical crisis. Here are some typical headlines taken from websites around the world:

"Man Drought Sees Shortage of Eligible Men as Women Struggle in Dating Game." (Michael 2014)

"Man Drought Leaves Many Lacking Romance." (Heather and Easton 2014)

"Hong Kong's Women Are Suffering a Man Drought." (Cox 2013)

"Amid a Growing Gender Imbalance, the Territory's Females Are Undergoing Drastic Measures—from Love Coaches to Liposuction—to Lure a Suitable Partner." (ibid.)

Or this quote from another article, its sensationalist tone backed up with numerical evidence:

Many single women looking for love could be out of luck, as a man drought turns severe. Census 2013 figures show the number of men to go around is at an all-time low—and it's especially grim for those of a prime marrying age. For every 100 women looking to snag a New Zealand chap aged between 25 and 49, on average about nine will miss out. And on the Kapiti Coast the chances are even lower, with only 82 men for every 100 women. If you're looking for a little older or younger gentleman the chances are better—51 per cent of the total population is female. (Heather and Easton 2014)

As the extracts illustrate, a cavalcade of demographers, sociologists, psychologists, economists, and lay experts of all sorts present themselves to analyze the data and estimate men's and women's chances of getting married. Often—as in Patton's letter—the experts evaluate the market value of single women, while warning them to hurry up because their market value is gradually declining. The study of singlehood—as in so many other social realms—reveals the extent to which the *tyranny of the market* (Bellah et al. 1985; Bourdieu 2003) and the commodification of social experiences infiltrate our lives. Indeed, a significant body of research proves these assertions, demonstrating how people draw upon the vision of the commodity markets to create, maintain, and renegotiate social ties with other people. Following this line of inquiry, scholars have shown, for example, that the commodification of intimacy and romantic love is congruent with the logic of rationalized market exchange (Hochschild 2003; Illouz 1997, 2007), and that procreation is becoming an increasingly commercialized process, in which eggs, sperms and embryos are treated as consumer goods.

Bauman's (2000, 2003, 2005, 2007) work provides another perspective from which to examine the ways in which personal relationships can be seen to correspond with capitalist and consumerist logic. For Bauman (2003)—who offers a pessimistic account of intimate relations today—the cultural configuration of *liquid love* operates within the unrestrained capitalist consumerist framework which frames personal relationships as goods and services. In Bauman's view, the consumerist logic forms frail-fluid, human bonds, or what he dubs *semi-detached, de-facto relationships*, made up of quick beginnings and quick endings. Online dating, in this sense, is tantamount to shopping for a partner—with no obligation to buy, and a generous return policy for dissatisfied customers (ibid., 45–46).

The analysis presented here seeks to explore how representations of single women's time in Israel is articulated and regulated by this commodity-imagery. It is important to stress that the single market place—which, in Bauman's words, operates within an unrestrained capitalist logic—is also attuned to collective temporal conventions. Putting it another way, I argue that the temporal component must be taken into consideration when one seeks to understand the power of this economic discourse in creating seemingly rational normative guidelines which are rarely disputed.

In order to understand the commodification of time, one must return to some of the classic scholarship in Time studies. This line of inquiry begins with E. P. Thompson's (1967) widely quoted study examining the transition from natural time to clock time, and the formation of clock time discipline in industrial capitalism. This significant transition, brought about by the industrial revolution, has transformed public culture and the sense of selfhood. As the vibrant historical and sociological literature on Time has demonstrated, time had become a standardized resource: to be calculated, allocated, saved, bargained, and controlled (Adam 1995, 26, 85). Or as Thompson puts it, "Time is now currency: it is not passed, but spent" (Thompson 1967, 61).

In this regard, Benjamin Franklin's famous aphorism, that "time is money," poignantly captures this social and historical juncture, within which a new temporal discipline looms large. Time calculations are considered as value-neutral objective frameworks, operating according to market demands. From this perspective, one can understand how perceptions of time as value and commodity play a central role in the calculative assessments of this consumer marketplace. Within industrial factories or outsourced call centers in India, time measurement is an instrument of power, producing undisputed temporal knowledge and temporal disciplines.

Returning to the single marketplace, I want to argue that these temporal customs are politically charged and implicated in regimes of power, which discipline single women through compliance with their fluctuating exchange value. Seen this way, ageist and sexist norms like those promoted in Patton's public letter are presented as temporal facts, merely reflecting market dynamics. In this context, this temporal economy works as a regulatory ideal which single women must comply with from an early age. This notion is exemplified in a column written by Lotti Kremba (a pseudonym), an Israeli single woman:

> [Addressing single women] Ha! At your age you suddenly remember that you want a husband and children? There you have it, you've lost! You missed the train; the train of the biological clock and social order, [you ignored] mortified parents, friends, and neighbors. But above all, you missed the train of men! Men—regardless of their age or looks— will always choose a fertile young woman with no wrinkles ... That's how it is. Nothing can help you now; you are off the market! A market in which there is no equality between the sexes. (Kremba 2009)

Lotti declares, cynically, that nothing can help single women above a certain age as they are "off the market." This is a market that privileges single men and younger women, she proposes. The temporal language of the market conveys a chauvinist logic which cannot be challenged or refuted. "There you have it, you've lost!" she states. It's a competitive market, in which one gains and loses in relation to one's value in the current exchange market. This is why single women should have known better, and should have looked for a husband when their youth still endowed them with a higher market value. Moreover, Lotti's tone conveys the message that it is the single woman's responsibility to stay alert and to practice constant self-scrutiny over her position in the market. Otherwise, she is doomed to become a "waste product."

As the above account clarifies, "missing the train" is an act which defies both biological and socially determined schedules. As an irrational actor, the single woman fails to acknowledge what is well known by her immediate surroundings. These messages have a firm tone: "But above all," "Nothing can help you," "That's how it is," "You are off the market." Moreover, as the writer states, the market does not adhere to feminist codes: there is no equality between the sexes, and there is nothing that the single woman can do about this. In this context, the single woman has no female agency and there is no possibility of resisting these oppressive gendered relations. Patton's Princeton letter deploys a similar rhetoric as quoted previously: "I'm going to say most women who are 35 are going to be looking for a man around the same age, or maybe a year or two older. So let's take the man of 36. He's quite happy to actually be with a woman 10 years younger" (Wallace 2014).

The power relations between men and women are configured here as neutral regulations of the temporal marketplace, one in which women's age is the currency by which their value is set. According to this logic, as single women age, their marketplace inferiority becomes an absolute fact, and thus they have limited options in finding a man their age "or maybe a year or two older."

The same ideas are expressed in Rachel Greenwald's (2004) best seller self-help book, *Finding a Husband After Thirty-Five: What I Learned at Harvard Business School*, translated into Hebrew and published in Israel. Greenwald's popular book propagates these messages by prescribing a business plan for single women over 35. In an interview with Dana Spector—a columnist with Israel's most popular weekly *Yedioth Ahronot* she urges single women to adopt a "realist" stance concerning their exchange value in the single marketplace:

> I'm not talking about admirable women who are happy being on their own … I'm talking about those who do want marriage and children. Women who want a husband and have reached this age, must begin to relate objectively to the market conditions which surround them. They have to stop dreaming that one day he will come along and must wake up and do something to find him. I call it a wake-up call, a necessary wake-up call. Men aged 30 and 40 plus approach me. A mass of them. They always tell me, "You must know a lot of women, can you introduce me to some? I really want to love someone." That's when I dissolve "Ohhhh, he is so cute," until they add in a nonchalant tone, "and by the way, I do not want to date anyone over the age of 30." It's cruel, it's terrible, but that's reality. I could say to him: "You're a pig and I refuse to help you," still I decided to do something else. (Spector 2004)

According to Greenwald, the single marketplace conditions are crystal clear: single men do not want to date women over thirty. This is why she views her book as a wakeup call to women who fail to realize how the laws of the market operate. The socially constructed inferiority of "over aged" women is not challenged, but rather reinforced by these claims.

Drawing on this line of analysis, we can see how this standardization of time is configured in an economy of ageist and sexist temporality. A woman's age signifies her exchange value and social worth. From a certain stage in her life course, she becomes easily disposable. Within these conditions, single women have to compete with each other over a sacred and a vital resource—the attention and time of single men. Single women who have been single for too long are accused of refusing to adapt to what are articulated as universal, biologically determined market-based rules. Hence, they fail to acknowledge their limited chances of survival in the market in the long run, and instead waste their time. According to this scenario, as in the workings of the consumer economy, single women are bound to be replaced by younger and "fresh products."

In her review of Greenwald's book, Dana Spector compliments Greenwald, describing her book as sincere and effective. Dana opines that anyone who implements Greenwald's tactics seriously will increase her chances of getting married:

> Single women who read Greenwald's book undergo a harsh reality check, which urges them to wake up and realize that their worst fears have become a reality: Yes, a single woman at this age is considered to be damaged goods, with a limited chance of getting married. In fact, this single woman has reached the September 11 of her singlehood; this

deadly Armageddon requires a re-organization. It sounds terrible, but Greenwald's frankness makes her book extremely effective for her target audience. Anyone who actually implements these focused military tactics may certainly well increase her chances of getting married. (Spector 2004)

Spector's tone recalls the praise garnered by Patton's letter, and could be summarized thus: "Finally, a courageous woman offers an effective solution to 'aging' single women." Only by adopting the strategy of the business plan can single women liberate themselves from the illusion that they still have time left. As in all the excerpts considered thus far, the temporal logic of the market is represented as a timeless truth, one which erases other social experiences. What's more, while these claims may appear harsh, they are—in Spector's view, at least—a necessary wakeup call.

Let's return to some of the messages underlined in Patton's letter. As she clarifies, when single women attend college, their market value soars; accordingly, they have more options to choose from and to be chosen. Following this logic, this is why they should put their youth to maximum value rather than waste their "best years." Singlehood in one's thirties can be nothing but gloomy, since men can—and will—date women ten years younger than them. Both Spektor and Patton, like so many other commentators express this kind of patriarchal concern for all the single women who fail to acknowledge the dynamics of the market place.

While some of these themes have been discussed in the preceding chapters, the focus of this chapter is to explore how these presumptions are commodified and absorbed into the language of temporal market exchange relations. As the textual analysis demonstrates, the statements and images through which single women are represented are imbued with age grading and age-based timetables, through which women are constantly objectified and evaluated. This is a pervasive instrument of social control, which confers normative standards and prescribes rhythms and mechanisms for inclusion and exclusion. The naturalized authority of time, coupled with the rhetoric of supply and demand infuse these schedules with potent discursive force and warrant the successful maintenance of a patriarchal, ageist temporal order. In other words, the temporal market rhetoric is presented as connected to the individualized decisions of the male consumer, and decontextualizes these decisions from its patriarchal tone.

This outlook is instilled with strong overtones of panic and blame. For instance, this tone is vividly illustrated when single women are accused of being "too selective," as well as in the popular demographic discourses discussed earlier. Marking single women as being too selective appears to be a global phenomenon, whereby the notion of selectiveness has come to be identified with the cultural figure of the urban, educated, and economically independent single woman (Lahad 2013). The accusation of selectiveness, with the attendant command to compromise, also carries with it the sensibilities associated with putting a mirror in front of single women and liberating them from their illusions. This tone, which casts doubt on their abilities to perceive reality, is also indicative of the infantilization process that single women are often subjected to. Injunctions such as "Grow up and learn how to compromise," refer to what is perceived as their inability to face the "facts of life," to recognize their fluctuating exchange status in the single market.

The truth-bearing quality of age and time, as discussed in previous chapters, endows these statements with the authority and aura of expertise. This discourse is arrayed in everyday parlance as well as through the plethora of dating experts all urging single women to hurry up and understand that they are running out of time. Otherwise, as they are repeatedly warned, they will turn into "aging spinsters" or "old maids," the figures that represent women excluded from the singlehood market, and thus exist out of society.

Single women as damaged goods

> Sometimes I feel like a horse at an auction. I allow myself to be examined by a stranger with a suspicious nickname. This man evaluates my weight and height, he compares my breasts in relation to others, examines my teeth. At the end of this examination, I receive a compliant, hum, not too eager, God forbid. In order to escape from the dating sites, you have to put yourself up for sale, as though you are the most attractive product: upload an excellent picture, provide remarkable information, and think about the consumer. (Levin 2006)

In the quote above, May Levin, a single woman, describes the process of self-commodification she underwent, during which she evaluated herself according to the objectifying gaze of potential men. Accordingly, her different body parts bear different values, through which she can be attractively displayed, marketed, and sold. Hence, in this marketplace one has to recognize one's fluctuating exchange value and do all one can to maximize this, in order to attract the attention of the male gaze. In this process, as feminist scholars have long emphasized (Bartky 1990), she becomes a self-policing subject, dependent upon the male gaze and his judgment.

Endorsing the temporal language of the market becomes a disciplinary apparatus, one which produces the docile subjects of the Foucauldian analysis. Its authority is accepted as legitimate, and becomes a potent form of self-surveillance. Our narrator above is aware of her position in this competitive market, and the subject positions ascribed to her in this particular discourse. In this account, she draws on the metaphor of the auction as representing the ways in which she is objectified: each of her body parts is measured and subjected to a market estimate.

The presence of the anonymous male consumer is also apparent in the next piece of advice, offered by Esta Brodsky-Kauffman, a dating coach who writes regularly for *nrg*, a popular Israeli online portal:

> You can bend the truth a little: It's okay to lie regarding your age by a year or two, to add some height and lose some weight. The objective is to present an attractive package in the eyes of the potential date. (Mendelman 2013)

As Esta explains, one's exchange value is dependent upon one's ability to present an "attractive package." Drawing from marketing tactics, the dating coach suggests that there is nothing wrong with bending the truth a little, accordingly making slight adjustments to one's age. "You have to think about the consumer," the expert emphasizes, "and present yourself as attractive merchandise" (ibid.). Following capitalist consumerist logic, the image is what counts; or as the expert puts it, the ability to present attractive merchandise or packaging. No surprise that many dating websites worldwide follow and endorse this logic by making space on their webpages for commercials touting hair removal, plastic surgery, diets, and professional photographers, who can maximize the candidate's capabilities in attracting the attention of male consumers and upholding market standards.

The acceptance of these scenarios yields another temporal metaphor, the belief in "the single woman's short shelf-life," as Louise (a pseudonym), a single woman writing for *Ynet*, declares:

> Let's face it, the shelf-life of singlehood is shorter than that of a tub of yogurt in the warm summer months; a single woman who allows herself to relax for too long on the couch will be lonelier than a voter for the Meretz party at Bar Ilan University. (Louise 2007)

In a different column mentioned also in Chapter 4, Merav Resnik, another *Ynet* columnist, coveys a similar message:

> Recently, I read an article about the different remedies that are supposed to cure the single disease for people above the age of thirty-five. While reading the talkbacks, I couldn't help but notice that one of them exclaimed: "Anyone that is single above the age of thirty is damaged goods." I smiled to myself; how lucky I am to be only thirty-two. My damage is considered to be light; I am safe for the time being. But the thought hasn't really disappeared. (Resnik 2007c)

The perception of the single woman's short shelf-life can be found in numerous jokes and popular sayings. One such joke was published in a 1986 *Newsweek* article (Salholz 1986) suggesting that a single, college-educated forty-year-old woman was more likely to die in a terrorist attack than to get married. In the 1980s, Japanese women were called "left over Christmas cake": just as no one wants to buy Christmas cakes after December 25, Japanese men are not interested in women over 25 (Dales 2014; Goldstein-Gidoni 2012; Nakano 2011). Although these sayings go back thirty years they are still present in contemporary discourse about singlehood in Israel and elsewhere.

For example, in a research study on single women in China, anthropologist Arianne Gaetano (2009) quotes a different joke circulating on the Internet: "A 20-year old woman is like a basketball, everyone scrambling for it; A 30-year old woman is like a ping pong ball, everyone hitting it back and forth; a 40-year old woman is like a soccer ball, everyone wanting to kick it; A 50-year old woman is like golf balls, the further away it is hit, the better" (ibid., 5–6).

The above accounts exemplify the ways in which the temporal language of market exchange has infiltrated personal relations, and the extent to which single women position themselves as marketable commodities—and in turn view single men as very selective consumers. It also conveys the threat of becoming unusable, the rejected objects of consumption. From a certain age, they cross the point from which there is no return. These views are echoed in the next web column, written by Hadas Friedman who refers here to her "damaged goods" label:

> Congratulations to me! I recently reached the age at which, according to some of the talkbacks, I can be tagged as damaged goods, or as an old single woman obsessed with marriage and children. This is the age at which I'm supposed to internalize the verdict upon me and understand that I'm being punished. If not now, I will be punished in the future for my selective, arrogant, and reckless behavior. This is what happens to a woman who has not married by this advanced age—she should understand that she herself has determined her own fate, and from now on she will remain alone. She should be aware that from now on, no man will ever want her. Why should he? He has the option of choosing younger and more beautiful women, and of course less selective ones. (Friedman 2009)

Hadas realizes that her status, as "damaged goods" or being called an old single woman obsessed with marriage and children, has deterministic consequences. She is now afflicted by the ultimate punishment: no man will ever want her. In other words, she is not marketable. The criteria are dependent upon principles of a temporal market which comes to view as a potent shared system of thought. No alternatives are possible, and resisting this temporal consumer logic is improbable. The threat of being consigned to waste is seen by Bauman as integral to the capitalist mode of consumption:

> Objects of consumption have a limited expectation of useful life, and once the limit has been passed they are unfit for consumption; since "being good for consumption" is the sole feature that defines their function, they are then unfit altogether—useless. Once unfit, they ought to be removed from the site of consuming life (consigned to biodegradation, incinerated, transferred into the care of waste-disposal companies) to clear it for other, still unused objects of consumption. (Bauman 2005, 9)

As indicated before, books like Greenwald's (2004) bestseller *Finding a Husband After Thirty-Five: Using What I Learned at Harvard Business School* accept these underlying socio-temporal presumptions as a given. Indeed, the aforementioned title could be paraphrased as *How to Find a Husband before Being Forcibly Ejected from the Marriage Market*. The age of thirty-five, in this case, signifies a state of emergency in which the single woman's condition has become a "9-1-1 situation," as Greenwald puts it (ibid., 3).

As Bauman states: "Life in the liquid modern world is a sinister version of the musical chairs game, played for real. The true stake in the race is (temporary) rescue from being excluded into the ranks of the destroyed and avoiding being consigned to waste" (Bauman 2005, 3). The fact that feminine singlehood extends through a longer time period than before does not imply that its boundaries can be extended endlessly; they are demarcated by rigid gendered age norms. Metaphors and jokes about the single woman's short shelf-life can be seen, in Bauman's words, as a "sinister version of the musical chairs game."

Continuing the discussion begun in previous chapters, it could be claimed that in the earlier stages of their singlehood, women have control of their time: consequently, they are "suitable for consumption," can be an active purchase to-be, while with the later ones they cannot be "kept in the store anymore." This is one of the reasons why women are socialized from an earlier age to do all in their power to avoid being consigned to waste in Bauman's terms. Their shelf-life is bounded by age limits, and therefore one should be aware of her expiration date or—as Bauman argues—when subjects are no longer fit for consumption.

However, instead of complying with this logic, I question the discursive parameters and the temporal measurements of this shelf-life. These measurements are ingrained with sexist and ageist beliefs. On this view, this temporal market language endows patriarchy with discursive force, and I argue that this should prompt us to think about how we can disconnect our thinking about women's expected life trajectory from these popular metaphors and seemingly undisputed laws of consumer capitalist circulation. These discursive templates are also incorporated in the related assumption that single women should be wary of wasting time, because their time is running out.

Wasting time/accumulating time

The injunction not to waste time is another temporal construct which reflects the vigorous temporal regulations to which single women are subjected. In this context, the prohibition not to waste time becomes a disciplinary norm, according to which single women are socialized into becoming their own watchful guardians and successful *time managers*. In his celebrated analysis of the Protestant work ethic, Max Weber observed that wasting time is considered to be "the first and, in principle, the deadliest sin" (Weber 1985, 157). This observation is related to the perception that time is both a valuable and an expiring resource. This principle is widespread in capitalist consumer society and, as we shall see, is present in many accounts discussing singlehood time.

As the following examples will attempt to demonstrate, the temporal imperative not to waste time places single women in a constant state of alertness. Meeting the right guy can happen at any moment. Ella Pe'er a single woman explains:

> I live in Tel Aviv and according to all social parameters I am a successful woman. I have a senior position, I earn a five-digit salary … usually I am surrounded by a lot of people and many desire my company. It is almost as if I'm surrounded by so many people so that I won't feel the loneliness anymore and I won't remember that I am alone in the world … it always seems that any second this might happen to me, yet reality proves differently. (Pe'er 2007)

In a different context, Rona Stern, another single woman, describes the pressure she is subjected to at work: "At work I am the only one who is not married; thus, every single guy entering the office must be scrutinized immediately: perhaps there is a chance that he will be the guy for me" (Stern 2007). In a similar vein, Esta, the dating coach, instructs single women: "Act like every interaction is your dream date; you have to be at your best most of the time to meet people out there" (Brodsky-Kauffman 2006a). In a different online advice column, she takes the case of Naomi as an example of wasting the best years of her life with the wrong man:

Naomi is a good looking thirty-two-year-old woman. After being in a relationship for many years with an emotionally unavailable man (an atrocious waste of time; indeed, she wasted the most beautiful years of her life), she decided to take action and to end the relationship. She began her session with me at the point in which she constantly found herself; at the beginning of a relationship which would not yield a thing … she's no longer twenty-two and doing what she has done so far has not really worked for her. (Brodsky-Kauffman 2008b)

In Naomi's case, as described above, she did not *invest her time* wisely; accordingly, her stock went down. Arguably, this bad investment did not lead to any productive results, such as finding a husband and having children. The prohibition of wasting time becomes a duty of the self. It is the single woman's responsibility to do all that is in her power to prevent herself from crossing the point of no return or in Bauman's (2005) terms leading *wasted lives.*

This *temporal awareness* configured into a constant state of alertness can also be found in the global speed dating trend also popular in Israel. One of the key assumptions underpinning speed dating is the scarcity of time. As one Israeli speed-dating agency, named *Speed-Date*, promises in an advertisement, their venture is intended for people who do not want "to waste time on pointless, drawn-out blind dates" (Speed-Date 2014). Consider, for example, the following questions in a different Israeli speed dating website named *Look4Love*:

Do you want to go out on more dates?
Are you tired of wasting time on long, unnecessary dates?
Are you interested in meeting more people in less time? (Look4Love 2015)

These queries illustrate some of the shared cultural understandings and social pressures in today's dating scene. A temporal reading of speed dating is interesting; it assures potential clients that it will acquaint them with the love of their life in just a few minutes. Speed dating, drawing on Bauman's (2003) analysis, can also be perceived as a new consumer practice in which potential partners become merchandise. In these market conditions, one must be quick and efficient. Consider questions such as: "Is this a serious relationship?" "Where is it heading?" "Does it have a future?" All these questions manifest an anxiety produced by the injunction of the desire to not waste time, but instead to invest in the "right" kind of relationship. Another best-selling self-help book, which in time also became a global box office hit screened also in Israel, was *He's Just Not That into You: The No-Excuses Truth to Understanding Guys* (Behrendt and Tuccillo 2009). The book's underlying message is that one should not waste time on "dead-end relationships."

In Israeli relationship terminology, there is a well-known phrase, called the *Yachasenu Le'an* conversation (the "Where is this relationship heading?" conversation). This is considered a crucial conversation that one is expected to conduct with a potential life partner, in order to determine the intentions of the partners involved. The underlying implication is the need to know if one should invest time in the relationship. The common understanding is that this kind of conversation is usually imbued with the prophetic power to determine if the couple should break up or stay together. Moreover, in Israeli culture this kind of conversation is readily associated with images of "needy" and "hysterical" single women, searching for the cues, confirmation, and reassurances for a future commitment on behalf of a future husband. Thus, the *Yachasenu Le'an* conversation reflects some of the prevailing cultural ideologies which hover over relationships that do not accord with linear productive temporal frameworks. In other words, relationships which do not accumulate to substantial value—getting married and having kids.

Seen this way, romantic relationships which are "going somewhere" are automatically configured as more meaningful and productive, and implicitly dismiss a vast range of alternative social relationships. According to this logic, short-term relationships, "flings," one-sided love affairs, and "platonic" relationships are categorized and graded in relation to "real" and "meaningful" ones: that is, long-term, committed romantic relationships.

What follows from the above is that the *temporal accumulative* dimension plays a crucial role in imagining familial relations. From this perspective, long-term relationships represent a positive accumulation of time. In fact, successful long-term relationships signify time invested wisely and productively. For example, in a different context, the *New York Times* reported a few years ago that the current marriage "crisis" has led to the declaration of a new public holiday in Russia (Rhodin 2008). In a press announcement, the State declared that couples married for more than twenty-five years would be awarded medals and declared ideal families in special ceremonies across Russia.

The way in which conjugal time is socially constructed as a positive accumulation of time also emerges in the next example, telling the story of an Israeli couple: "Ruby and Natalie, so it seems, have not wasted their time and they have accomplished the goals they have laid out for themselves" (Reinstein 2009). This tone of appreciation is expressed by a *Ynet* writer in a special section of its website entitled "Couple of the Week," in which an Israeli couple share their story with the portal's readers: "We both subscribed to *JDate* in order to look for a serious partner and create a family ... after three weeks of dating we felt that this was it ... after a month-and-a-half Ruby proposed" (ibid.). The column ends with the celebratory statement: "Today, a year from their first date, they are on their way to starting their family" (ibid.). This linear model of couplehood organizes time as a sequential process of steps and stages. Like many similar narratives, the happy couple are praised for not wasting time and acting in accordance with societal timetables. Further, by focusing on their goal-oriented pathways they have successfully utilized appropriate time-measurement techniques and norms. A similar story is the story of Galit and Tal, also appearing in the "Couple of the Week" section of *Ynet*. Both partners are in their thirties; after one week of knowing each other they decided to live together and two weeks later decided to get married (Farkol 2007).

In the preceding chapters, I emphasized the temporal imperative of successful timing. According to the pervasive cultural scripts in earlier stages of singlehood, many women are encouraged to experiment and to engage in different relationships which would not necessarily lead to marriage and children. This highly recommended experimentation is considered a necessary step toward enriching one's personality, and is even regarded as an important basis for one's future and for building a "serious relationship." On the other hand, moving in together after a week once you are in your thirties is perceived as a wise and recommended decision. One's age is figured as a primary coordinate of meaning and evaluation, as well as a crucial focal point for setting priorities, planning life goals, and estimating successes and failures.

One more related example can be found in another one of Esta's columns, when the dating advisor addresses a twenty-seven-year-old woman who has never been in a romantic relationship:

> You are rightly concerned. Statistically, someone who has never been in a relationship and has passed the age of twenty-five has a serious problem. Your chances of being in one in the near future are increasingly diminishing. According to a wealth of research, people who have never experienced a mutual relationship lack the basic qualifications needed for maintaining a conjugal relationship ... They lose the little things which make someone capable of being in a relationship ... With such a resume much luck will be required. (Brodsky-Kauffman 2007b)

As evidenced in this column, late singlehood is ascribed with negative and pathological social connotations. The longer you are single, the harder it is to get out of singlehood, to become a non-single. Singlehood becomes part of your identity, constituting one as a loner, diminishing one's market value, forming bad habits and, as Esta claims, causing one to "lose the little things which make someone capable of being in a relationship." Thus, being single for too long leaves one with emotional and behavioral deficiencies requiring different forms of "rehabilitation" as a precondition for re-entry into the marriage market.

Elsewhere, Esta declares: "There is a fact that dating sites want to keep hidden—anyone who does not succeed finding a partner within 3 months, will remain there for years" (Mendelman 2013). In another column, "Trapped in the Net," the dating expert quotes what she dubs as the terrible statistic that most dating websites would be reluctant to uncover: the longer a person frequents a dating website, the more their chances of meeting someone diminishes:

> Do you know, there are some very alarming statistics which online dating sites will do everything in their power not to reveal: the longer a person spends time on dating sites, the less chances one has to meet someone. Yes, yes, this is a real statistics. And why is it so true? Not because of a bad spell, but due to the behaviors people adopt on these dating sites. Don't get me wrong: I'm all for dating sites. They certainly create opportunities to meet people who in other circumstances you cannot meet but the use of them can turn out to be dangerous if it is not done right. (Brodsky-Kauffman 2009)

So, according to Brodsky-Kauffman, the longer someone is single the less chance they have to *unsingle* themselves. The above analysis exemplifies how heteronormative modalities of time acquire coherence and normativity. As Kerry Daly suggests, through the moral economy of time one has to be a good manager of one's time and successfully meet deadlines in which certain amounts of activity have to be condensed into a specified period of time (Daly 1996, 86). The single woman should use her time prudently otherwise she loses her ability to act, both in the present and future. She becomes trapped in the net and the pathologies of long-term singlehood. This is reminiscent of a new pathology inscribed to singlehood: chronic singlehood.

A few years ago, during a class about the medicalization of feminine singlehood, a student drew my attention to a new term adopted by some Israeli journalists and experts: *chronic singlehood*. In an article published on the *nrg* portal, a journalist describes the following method which would supposedly "rescue singles from their pathology":

> Dr. Ora Golan offers another method, which has nothing to do with statistics. She is a Doctor of Chiropody by profession, and offers to release every eternal single man and woman from years of loneliness. This is done through a series of sessions of 20 minutes … Golan reports up to 80 percent success with her method. (Dagan 2010)

It is beyond the scope of the discussion of this chapter to present an extensive critique of the medicalized and therapeutic tones of the term, yet I do think its temporal dimensions merit our attention here. The term "chronic" signifies both the negative accumulation of time as well as a loss of control and autonomy. In this context, I emphasize once again the connections between time and agency. Terms such as "chronic singlehood" or "the single woman's short shelf-life" come to designate *a loss of agency* and a vastly diminished capacity to act and determine one's life trajectory. In this sense, women are controlled by their illness and or diminishing exchange value in the single market place.

These medicalized and therapeutic rhetorics become part of a set of disciplinary techniques modeled on the competitive marketplace. As the textual analysis here demonstrates, the circulation of models, metaphors, and phrases which draw from this language is abundant, and interacts with economic and temporal deterministic assumptions about the possible life trajectories of single women.

This deterministic tone also explains why one rarely encounter texts that run counter to this temporal economic logic. An example which offers an alternative to this form of thinking can be found in the next column:

> I often hear how people only want a serious relationship or a relationship that will lead to a wedding … it sometimes seems as if we live in the realm of what will happen in the future only, and consequentially we limit ourselves with lists and requirements that we have invented … We check, select, and examine and we miss out not only on the future but what could also be the future. Perhaps we should learn and perceive life as a holiday; a journey made up of summer flings. As not just about what we could get out of it; not just about the race, but what we feel here and now. What will happen? Will we fall in love? Will our hearts break? Damn! It would at least remind us that our hearts exist. (Friedman 2007)

Hadas, the writer of the last column, lucidly critiques the current culture climate in which personal relationships are now constituted. The primacy of long-term relationships, it seems, rules out any possible alternatives. To put this another way, she points out the mere fact that social life is made up of a plurality of social interactions, yet

these discursive tenets cast off relationships which fail to conform to linear and productive paradigms, setting rigid limits on what is considered as meaningful and valuable.
According to this mindset, short-term relationships and long-term relationships which
have not necessarily resulted in successful marriages are regarded as an utter waste of
time. Common behavioral standards which attempt to distinguish between wasting
time and investing time fail to consider the affection, magic, interest, and meaning
often found in relationships which do not necessarily result in marriage and children.

To conclude, this chapter has addressed the ways in which the commodified language of time shapes the popular perceptions of female singlehood. My analysis here
has also attempted to shed light on some of key questions which preoccupy sociology,
cultural studies, and feminist theory. Among these is the objectification and commodification of women. Age branding in this case plays a significant role in determining
women's exchange value, and women's sense of agency and social worth. The unquestioned adherence to these beliefs provokes women's anxieties throughout their life
course: "Can I still participate in this marketplace?" "What is my current worth?"
"When will I be disqualified and excluded from this commerce or become 'damaged
goods'?" The analysis presented in this chapter should prompt us to create alternative
discourses to the *commodification of single women's schedules*. Within this context, I
stress the need for alternative ways of thinking about singlehood, disconnecting singlehood from normalized concepts of market logic, exchange value, and single women's
capacity to be "sold" and "exchanged."

The texts analyzed in this chapter voice the current devastating effects of patriarchy.
In this manner we can see how patriarchy overlaps with other systems of oppression,
namely singlism and ageism. Thus, rethinking time norms and schedules should be
located in a broader framework of gender forms of oppression and vice versa. That is,
feminist resistance should pay attention to the ways in which temporality is exercised
to discipline and normalize the female body, punishing who ever cannot comply with
these fixed rules. Thinking beyond the conjugal imaginary poses such an alternative
which will be further discussed in the last chapter of this book.

6

Taking a break

> In one day in the modern world, everybody does more or less the same thing at more less the same time, but each person is really alone in doing it. (Lefebvre 2004, 75)

The life courses of single women are characterized by shifting schedules, rhythms, breaks, and returns. However, as they grow older the common perception is that their time becomes a sacred and a highly limited resource. This now limited timeframe requires new temporal planning and, accordingly, they are expected to do everything possible towards achieving their overriding life objective: un-singling themselves (DePaulo 2006). At this point in time, the pressing need to take action and speed up the search for a suitable partner becomes evident. Against this background, an extended break from conjugal relationships or from "looking for the one" becomes bestowed with clear temporal limits. These temporal perceptions do not tolerate any pauses, especially given that the clock will keep on ticking, entrenched in a general cultural climate of time urgency, what Negra (2009) describes as *time panic*. Consequently, single women must take all possible steps to avoid unnecessary pauses and to evade distractions or diversions from the relentless tempo generated by the search for "Mr. Right."

In this chapter, I examine some of the background discourses that create these beliefs, and evaluate their ensuing consequences. This, I think, presents an opportunity to examine how temporal constructs such as a *timeout*, *time on hold*, or *frozen time* are contingent, situated in contexts which are relational to single women's positions in the life course. Developing this theme further, the discussion that follows will unpack concepts such as *taking a timeout*, *breaks*, *speed*, and *mobility*, and will problematize the cultural script through which single women are perceived as having *their time on hold* and/or *being frozen* in time. These socially defined temporal schemes not only prescribe rigid models of temporal rhythms, but reinforce what are considered as natural and inevitable modes of being.

Many of the extracts analyzed in this chapter will demonstrate how single women internalize such beliefs, by negotiating, conforming to, and resisting them. Some of the women are viewed, and perceive themselves, as being "stuck"; others resist these norms and claim their temporal agency. In the following reflections, the concept of time is unpacked as being both dynamic and static: a resource that should be acted upon, but at the same time an entity beyond anyone's control. Single women are expected to act between these multiple poles of time.

In what follows, I seek to develop a new framework for rethinking these temporal templates, and to scrutinize the effects that these templates have on one's temporal self. By also engaging with scholarship seeking to queer time, I call for a re-articulation of prevailing normative narratives of time, particularly in relation to its rhythms, demarcations, and restraints. Following Halberstam's suggestion of examining the possibilities of leaving "temporal frames of bourgeois reproduction and family, longevity, risk/safety, and inheritance" (Halberstam 2005, 6), I will also examine how late singlehood offers such a possibility.

A timeout

In his beautiful essay "The Adventure," Georg Simmel (1997) grasps timeout as a platform upon which people detach themselves from collective rhythms and causalities. From this viewpoint, the adventure is seen as a positive timeout, embedded with intensity and excitement, a time which lets us "feel life in just this instance" (ibid., 230). Simmel explains that this experience is possible because adventures have unique qualities, which enable them to be liberated from the before and after. Nevertheless, even adventures are clearly demarcated by temporal limits. According to Simmel, "We ascribe to an adventure a beginning and an end much sharper than those to be discovered in the other forms of our experiences … An action torn out of the inclusive context of life and that simultaneously the whole strength and intensity of life stream into it" (ibid., 222).

Simmel's observations concerning the adventure uncover a unique temporality, a mode of being which is made possible due to its clear boundaries, demarcating its beginnings and its ends, the before and the after. I have turned to Simmel's essay because singlehood is often perceived as a short-term adventure, a break, a timeout, a time for experiencing, dating different people, travelling, soul-searching, and living on one's own. Yet this adventure, as we will see, can only temporarily be ascribed with a beginning and an end. Thus, I find the construct of the timeout as particularly enriching for our analysis, especially as it entails possibilities for resistance.

In a broad sense, therefore, a timeout can be perceived as a time within which individuals can relax and play, as they are temporarily released from socially dictated temporal frameworks. While on holiday, for example, one is given the chance to break away from clock discipline and mundane behavioral patterns.[1] In a similar vein, the popular, globally distributed magazine *Time Out*, for example, highlights this perception in its very name, proposing that a *timeout* is not just a time to be *out* but might also suggest it to represent leisure time, taking a timeout from work.

The length of a timeout can vary, from the brief two-minute pause during a basketball game to a break that might last weeks or years; it could be spontaneous or well-planned, chosen or imposed. However, timeout as a concept should be viewed in relational and situated terms. This perspective was employed by Ÿian (2004), in his fascinating work on unemployment among young people in Norway. Ÿian distinguished between two different types of unemployed youth: those considered as taking a *timeout*, and those who *dropout* from the linear trajectory altogether.

While the latter denotes a young, working-class school dropout, the former describes an upper-middle-class young man, one who took a break from his education and career trajectory. Each case signifies different subjects, produced by discourses of temporal linearity. While in the dropout case the young unemployed man is cut off—perhaps permanently—from the linear path, the timeout case represents a temporary departure from linearity, the implication being that he will eventually, and at a time of his choosing, re-join the linear trajectory. This particular timeout mode reflects a continuation between present and future, while the other stresses discontinuity (ibid.).

Social interpretations of the concept of a timeout can additionally express relations between one's self-identity and social constructions of linear time. In pursing Ÿian's theoretical proposition, I argue that the parameters of age and gender play a crucial role in differentiating between constructions of early singlehood as a legitimate timeout from the linear trajectory, and singlehood at a later age, where it is perceived as "jumping off the train" or dropping out altogether (ibid., 179). As per Ÿian's contention, two different modes of temporalization are presented here, conveying different subject positions and life trajectories.

The next passage, written by Dana Sa'ar, a single woman writing on the *Ynet* portal, exemplifies how timeout is contingent upon gendered age norms, and is regulated by societal timing:

> People will think you're strange if you say that you are taking timeout. Who needs a timeout? What do you mean by a timeout? Who needs a break from sex and love? What happened? What exactly aren't you telling us? We always knew that something was wrong with her! (Sa'ar 2007)

Dana Sa'ar expresses here the suspicious attitudes she encountered when she announced that she was taking a break. Who would possibly wish to take break from love and sex? The claim that one needs a break triggers the suspicion and criticism that single women regularly encounter in their everyday lives. Dana's quest is another indication that "something is wrong with her." In this instance, taking a break means detaching oneself from the heteronormative dictum that one must be on the perpetual, frenzied lookout for a husband.

These commonsensical perceptions are an apt example of the temporal regimentation of social life. According to this view, the non-stop journey during which one date follows another is regarded as a productive and meaningful temporal trajectory. To detach oneself from the constant yearning and search for a potential husband is inconceivable. Dana further describes her experience.

> Recently, I ended a long relationship. During the first few weeks, my friends shared my sorrow, bringing me food as though I had just been widowed. But as time went by, they became less concerned, more bored; their encouragement was now formulated in terms like "It's time to get back on the market," "It's time to get back on the horse," and other phrases that made me feel like I was up for sale. Believe it or not, this does not suit me

... I don't want to enter a committed relationship no matter what, just so that I will not
be, God forbid, a single woman ... what's wrong with a little quiet time for myself? (ibid.)

Dana Davidovitz, another writer for the *Ynet* portal, narrates a similar story:

At the age of 30-plus, and after a series of disappointing and tedious dates, I have decided
to take a break, a sabbatical, a fast from dating, whatever you wish to call it. Nonetheless, in the terms of Israeli society, this is considered a hubristic decision. Who do I
think I am, how dare I leave the race to the *Hupa* [bridal canopy]? Although I have
clearly stated that I do not wish to be fixed up with any man, be he single, divorced with
children or widowed, my close family and friends keep handing me phone numbers of
men who are potentially marriage material. In addition to that, after a week they check
up on me to make sure that I called him. How is it that in Israel of 2010, a single woman
who dares to take a break from the tedious search for a date is so harshly criticized?
(Davidovitz 2010)

Dana Davidovitz points out that her wish for a pause from the dating world is interpreted by her environment as unjustifiable hubris. In a similar yet different vein, these
sentiments echo Karen Stein's (2012) study on the temporal experiences of vacations,
which observed that taking a vacation for too long can be viewed by others as indul-
gent. Dana comments that in Israeli society, her behavior is construed as arrogant, and
further reflects that as a thirty-plus single woman, she has no option but to join the
collective "race to the *Hupa*." These sets of beliefs are also consistent with the current
post-feminist rhetoric in Israel, one which urges women to return to their heteronormative life trajectories and traditional feminine roles.[2] As a result, marriage and parenthood are illustrated as undisputed life goals, which do not allow for senseless pauses
away from the pressurized search for Mr. Right, no matter how tedious the search
might turn out to be. Seen this way, timeouts are considered as nonproductive and
meaningless time, during which single women have "failed" to progress towards realizing their prescribed life goals.

It could be argued then that an overly extended timeout bears the risk of distancing
the subject from the future, or having no future at all. Both columnists regard time as
a resource which they can take and own. I suggest that the very expression "taking a
break" can also express the desire to take control of time, attempting to prescribe the
subject's own pace in a collectively determined timetable. This in part might explain
why their autonomous claim for time prompts so much criticism, as it both defies
conventional socio-temporal norms and asserts a sovereign selfhood which does not
conform to these prearranged rhythms. The common explanation that many single
women hear is that taking a break is a *temporal privilege* that they can no longer afford.

In her attempt to restructure her own time, Dana Davidovitz draws on various
metaphors:

Why do I need to fast or take a sabbatical now? Here is my answer: only those in the
dating world can testify as to how difficult this ongoing search for something real is. Only
those who are in the midst of the race and really want a relationship can understand how
difficult it is at the age of 30-plus, to know someone and then discover that he is someone
else ... Only someone who has been searching for so many years can understand what

it feels like to be disappointed, to feel that you have failed and to know that despite everything that from this train, you do not descend on your own but only as a paired unit.

Indeed, this is what you really want and you can't let go. This sabbatical for me is a post breakup period … It is the time for me to gather my strength and look what is right for me. At this time I have no place for a new man. I am not a robot or an athlete in a marathon. This is the time for me to piece together my broken heart. It's a difficult time as it is, and all the advice [she addresses this to her pressuring environment] that you are giving me just makes it harder. I'm going out. I'll be back soon, but until then please give me a break. (Davidovitz 2010)

The writer configures the dating world here as a race, a marathon, and as a train. In subsequent chapters, I have discussed the metaphors of the train and the biological clock, which allude to the social pressures and accelerated pace to which many single women are expected to comply. In the above extract, the metaphors of time are significant discursive resources, which assist Dana Davidovitz in expressing and communicating the social pressures she experiences. They also enable her to break away (even temporally, as she herself admits) from constraining time pressures. I stress again the power of temporal metaphors, by quoting Ramón Torre's (2007) significant observation. For Torre, time metaphors are:

Ways of speaking, conceptualizing and experiencing [time], it is no less true that these ways are also (or end up being) ways of acting or doing. I therefore assume that the way in which the agents conceive of and speak of the world is also a way of shaping it. (ibid., 160)

These temporalized metaphors, as Torre suggests, are not merely ways of experiencing time, but actually a way of shaping time and re-conceptualizing what is conventionally considered as an interruption of the linear sequential flow of time. Indeed, the two columns accentuate the temporal boundaries of the socially legitimate timeframes for entrance, exit, and re-entrance to the linear flow. As one of the columnists pleads: "Give me a break."

The reflections quoted above are a useful focus for a discussion of the limited temporal tolerance towards single women who "use up their break time." To put it another way, a timeout or a pause between one relationship and another is counted, measured, and regulated by socio-temporal norms and their ensuing rhythms. Accordingly, the levels of approval, empathy, and support accorded to the single woman are determined by these temporal dictates. Both writers oppose and accommodate these dominant temporal frameworks. As Dana Sa'ar (2007) remarks, after a while her friends' empathy transforms into social pressures, articulated by instructions such as "get back on track," "get back on the horse," or "get back to the market."

If we draw on Ÿian's (2004) line of analysis, the writer's timeout is now re-interpreted as a preparation for her expected reincorporation onto the linear path. The legitimate time granted for her to "get over" her ex-boyfriend, along with fluctuating social expressions of empathy, are limited by these temporal social rhythms. The time has come to nudge the single woman forward, before her timeout turns into a dropout. This is exemplified by Dana Sa'ar's account:

When I told my friends that I want some quality time to myself, I got the message that the best way to get over someone is to meet a new man … and just by chance, they had this amazing person to fix me up with; well, perhaps not amazing but really nice. Apparently, now I'm no longer allowed to be picky. We all saw what came out of my crazy standards [referring to her last relationship]: neither a wedding nor children. Now is the time to settle down and compromise, and at my age you can't wait much. Moreover, all of a sudden my mother began to look at me as though I had murdered her future grandchildren … I have one or two years before I become too old; I can still make it … All I want is a little break … I don't know where this fear of breaks comes from; at school we all used to like them. (Sa'ar 2007)

As we can see, status transitions are inextricably bound to pressing social and temporal norms. Dana Sa'ar's timeout as a temporal interval is socially legitimate only insofar as it conforms to specific temporal norms and gendered age-based limits. During the first few weeks after her breakup, the writer is still positioned within the confines of heteronormative culture. The breakup is understood within the temporal order of conjugal and family life. Her timeout following the break up is considered as legitimate, highly recommended, and indeed "entitles" her to social support. Yet, as time passes and Dana moves further away from the world of couples and towards the world of singles, these levels of tolerance, empathy, and social support towards her single status gradually reduce. The more she distances herself from the agreed-upon and expected teleological journey, the less her social surroundings support her.

By the same token, her status transition is tolerable as long as it is understood as a temporary phase. When the temporary threatens to become permanent, the fragile social order is de-stabilized. As she comments, we all liked breaks in school; indeed, we long for breaks from work or other mundane routines. But taking a break from relationships for too long and at a certain age is far too risky, and therefore inexcusable. Timeout, taking a break, getting away, or taking one's time: these are all encouraged and considered to be legitimate in certain settings and at certain times. The realm of personal relationships has no explicit, institutionalized norms determining the right length of the break (unlike a vacation or any other fixed time period away from work, for example). Even so, at a certain tentative, yet socially agreed upon point in time, the bells all ring out vigorously, urging the single woman to return to class before she is thrown out.

Dana Sa'ar frames her timeout as legitimate: "one or two years, before I become too old." However, both writers are accused of not adhering to heteronormative social schedules of time, of ignoring time and not attending to its norms and requirements. Their quest for a break is a claim for a different temporality and rhythm. In this way, they defy dominant themes of time-use, normative rhythms and schedules.

In this we might consider their quest for a break as a form of *queering time*. Scholars like Tom Boellstorff (2007) and Judith Halberstam (2005) reformulate such linear teleological trajectories by suggesting temporal modes which do not conform to heteronormative and kinship paradigms. A timeout without clear and rigid bounds could be seen to fit such a temporal mode, one which conveys a non-purposeful and a non-progressive movement: dropping out of time in Ÿian's (2004) terms. In so doing, they emphasize that their timeout is a temporary one, a pause for recuperation before rejoining the dating race. As Dana Davidovitz exclaims, "I am out, I will be back soon" (Davidovitz 2010); standing outside dominant linear narratives can never be anything other than temporary.

Thus customary familial reproductive schedules can only be temporarily suspended. Otherwise, their timeout will soon become a dropout. While Ÿian has examined, in relation to employment, how one's class membership marks one as a timeout and the other as a dropout, in the case of single women, the axes of age and gender are important parameters which can lead them to a futureless life track. This becomes particularly evident when marriage comes to represent progress, civility, and futurity (Boellstorff 2007; Warner 1999).

A different example of how an overly extended timeout transgresses socially prescribed boundaries is articulated by Esta Brodsky-Kauffman, *nrg*'s dating advisor, when she writes about the increasing number of single persons:

> The result is … people come to me at the age of forty and want to marry or are looking for a substantial or a meaningful relationship, but according to what experience? … If you've had a break of a few years since your last relationship, there is a serious gap which you have to overcome … You're a little bit bitter, perhaps a bit frustrated … for the young there are fewer criteria … they are not accustomed to being on their own. (Brodsky-Kauffman 2006b)

The dating coach's complaint alludes to the unjustified timeouts taken by forty-year-old singles. She grasps the break as a "serious gap," one which will yield severe repercussions, as it seriously damages the single person's ability to engage in a long-term relationship. However, her understanding of this break is filtered through pervasive cultural beliefs of time. Young single women are privileged with the possibility of being able to take timeout: they still have time and can control time. In this account, one's age, gender, and relationship status are crucial parameters for evaluation.

Time on hold

The word *timeout* represents a double temporal motion: that the single woman's time is on hold on the one hand, while her peer group is "advancing" forward to marriage and familial life. The idea of a timeout within this formulation can convey the message that one's life cannot begin and is devoid of meaning, as shown in the next passage written by Shirli Malachy, a single woman:

> I am missing out on my life. I'm pretty, clever, self-aware and a laid back kind of woman … But despite it all I am missing out on my life. I'm living my life waiting for something to happen. I have put my life on hold and I wait, wait, and wait. I am waiting because in addition to all the above-mentioned qualities, I am a single 34-year-old woman. I'm living my life, and feel [that it can begin] "only when this [finding a boyfriend] happens." I'm just letting my life go by, counting the days until the right partner arrives … My life at the moment is devoid of any meaning … My ex-boyfriend wrote to me that now that he has gotten married he feels he can live his life. (Malachy 2010)

Shirli feels she is missing out and not living the life she ought to be living. She explains that she is waiting, counting the days until her real life can begin. She attributes the reason that her life is on hold to the fact there is no man in her life. As opposed to her ex-boyfriend, who claims that his real life began once he got married, she is "missing out on her life." She also adds that she experiences her life as devoid of meaning, and as a result she has nothing to look forward to or anything worth living for.

In her study of people living with HIV-positive diagnoses, Michele Davies (1997) contends that time is a platform for how and from where we live our lives. In this regard, one's orientation towards time is crucial to one's actions and behavior, and as such it is significant to one's understanding of human existence (ibid., 562). Drawing on Blaise Pascal, Davies argues that our dominant temporal orientation is that which predominantly projects us into the future, as we care little for the present. Quoting Pascal's book from 1889:

> Man cares nothing for the present, anticipating the future, finding it too slow in coming, as if one could make it come faster. Or calls back the past, to stop its rapid flight … so frivolous are we that we dream of the days which are not, and pass by without reflection those which alone exist … The present generally gives us pain; one conceals it from one's sight because it afflicts one, or if it is pleasant there is regret to see it vanishing away. The present is never our end; the past and the future are our means, the future alone is our end. Thus, we never live, but hope to live. (Quoted in Michele Davies 1997, 562)

Such a view of the present and future is evidenced in many of the texts analyzed, in which the present is construed as empty and devoid of meaning. In the column mentioned above, Shirli reflects upon what she terms as her life being on hold:

> Above everything else, I sense that there is a huge sign: my life is on hold until the right guy arrives. … Yes it's true, I do have friends and family but at the end of the day each and every one of them leads their own lives … What can you do? (Malachy 2010)

It is worth noting that Shirli comments that it is impossible for her to enjoy the present as everything seems meaningless without a male partner. Her account illustrates once again the power of heteronormative and familial temporality. In his discussion of the contemporary construction of family life today, Brian Heaphy (2011) claims that the family is a powerful story that cultures tell about the relationships that matter most. Thus, it gives priority to the family form over other relationships such as friendships, community, partnerships and so on. According to Heaphy, "Family is so 'naturalised' and taken for granted that its discursive and fictive nature very easily slips away from view. Its effectiveness as a form of relational governance is evidenced in how difficult it is for relational practices and displays to escape being viewed through the family frame: as family or not" (ibid., 34).

The effectiveness of this form of governance is present in most of the reflections written about female singlehood. As such, other relationships are perceived as secondary, viewed, as Heaphy so aptly claims, through the normative family and couple-oriented frame.

Frozen time

In her study of women's experiences of infertility, Becker describes the temporal experiences of some of her interviewees as a "culturally propelled sense of motion through time [which] had stopped" (Becker 1994, 396). Below, I quote from one of Becker's interviews:

> I have had my life on hold for so many years not thinking that I'll be pregnant … Intellectually it is almost inconceivable to me how you can contain or put your life on hold like that for so long and not go bananas but I've done it. I've lived it for over five years. It is probably the most frustrating aspect of infertility in my mind. It's horrible living in limbo. I think it affects your every waking moment, thinking about what you should be doing, what you could be doing, and what you want to do and yet you can't. (ibid., 397)

This sense of time, which has either stopped or sped up while one's life is on hold is also manifested in common representations of single women as "stuck," "not moving ahead," or "waiting for the one." Merav Resnik, a single woman and a columnist, has described this experience by using the metaphor of the "dating carousel," within which she describes herself as feeling as though she is moving again and again in a circular, purposeless motion. In her column she refers to this as a "rebound period," which she explains further:

> I am beginning to understand that this time period is characterized by a horrifying impotence. You want to be in this place called "onward"; you see it, you sense it, smell it, you can touch it on the tip of your fingers, but somehow you wake up in the same dammed spot. (Resnik 2006)

The same experiences are unfolded by Orit Gal:

> The holidays are a terrible time for the lonely. I know that pretty soon, people will call to wish me a Happy New Year, people whom I have not spoken to for several months. And then, eventually, the question about my singlehood will pop up. And again, I will have to explain that this year, again, I don't have a partner and that nothing has changed since last year. (Gal 2007)

As will be elaborated upon further in the next chapters, Valentine's Day, New Year's Eve, holidays, and weekends are often perceived as moments of crisis, partially because they function as symbolic time markers accentuating the interplay of personal stasis and the continued flow of time. In this manner, nothing is happening; one is *still* single and without anyone to kiss on Valentine's Day or New Year's Eve.

This experience of being stuck also draws on another temporal expression of becoming frozen, detached from one's past and future. Moriah Shalom, for example, describes the common features of a "frozen single person" when she writes about the man she is currently dating:

> When he emptied his frozen refrigerator and then defrosted it after it hadn't been touched for months … I understood that he is exactly like me; a frozen single person that once, a long time ago, had a life. But somehow, from one unsuccessful date to another, between love affairs that lasted a month or two, this kind of life was lost. (Shalom 2006)

In another text, "You Freeze in Fear and Then You Miss the Train," Merav also describes the experience of the frozen position in relation to the experience of singlehood:

> Totally frozen, we move neither to the right nor to the left … we are just stuck in place out of fear. We don't talk, we don't disclose, we are just silent. Everything is bubbling inside us and we stand in our place … [We] will leave decisions in the hands of fate … We're afraid to take responsibility for our actions, wills, feelings, desires; afraid to take chances and gamble. (Resnik 2007a)

Frozenness is congruent with passivity, lack of initiative, being bound by one's fears and bad habits. The single woman is represented as trapped within a repetitive temporal routine of stasis and inactivity. Kathy Charmaz (1997) has interestingly described this temporal phenomenon as a *slowed down present*—a perception of time which moves slowly while one seems to stand in place, resigned to one's fate.

In this sense, the image of a refrigerator in need of defrosting is reminiscent of the slow and gradual process by which a slowed-down present turns into a time on hold, or frozen time. The single woman, in this sense, is configured as a figure of stillness, left behind because she "failed" to catch the train in time. According to this social temporal imagery, the single woman has to be worked upon and put back into the right linear trajectory.

It is interesting to note that the technological innovations of freezing women's eggs today provides many women with the ability to claim their *temporal agency* and postpone giving birth to a later stage in life. In 2011, the Israeli Ministry of Health announced that women between the ages of 31 and 40 who wished to freeze their eggs for non-medical reasons could do so, thus allowing them the chance of giving birth later in life (Levi 2011). This decision led to an impassioned public debate, opponents of the use of this technology claiming that such innovations could promote the dangerous illusion that women could have children at any age.[3] It is beyond the scope of this discussion to explore the medical, ethical, and sociological aspects of assisted fertility technology in detail, but it is worth noting that this technology is being taken into consideration by society, and I presume its diverse effects will have some effect on how temporal schedules are, and will be, imagined in the future.[4]

Immobile subjects

Speed is a dominant aspect of contemporary culture, seeping from the domain of work into other aspects of life like family life and leisure patterns of sociation. Speed is associated with decisiveness, time management, and punctuality, and is perceived to be a celebration of human power. Living in the "meantime," by way of contrast, denotes stasis, indecisiveness, and passivity, all of which are considered to be reprehensible qualities in our speed-driven culture.

Charmaz (1997) writes that one of the predominant temporal experiences of people suffering from chronic illness is one of being held in abeyance. Time on hold is portrayed by Charmaz as an experience characterized by agonized waiting, the present and future unsettled and undetermined but might yet lead one towards a disastrous finale. In this respect, as Charmaz observes, the self becomes temporary, as it experiences that the future cannot begin. As one of her interviewees comments, for example: "It suddenly dawned on me [that] I really don't have any goals … nothing concrete. And I have been putting myself on hold … I'd feel like my life is aimless" (ibid., 191).

In another interview, the interviewee describes "time like a rope around me—when I feel optimistic I let it out, the time just unfolds. When I'm feeling pessimistic the rope is tight" (ibid., 190–191). Or, "The dreaded future engulfs the present self. Likely, someone puts his or herself on hold. The self-experienced now, be it the grumpy, fearful, martyred, apathetic, or withdrawn self, becomes a temporary self" (ibid., 33).

Similar themes surface in Reith's (1999) work on the temporal experiences of ex-drug addicts. Her analysis notes that many of them describe their past as lifeless and static. As we have seen, these experiences of non-movement and of standing still, the sense that one's life has stopped and that the future is blocked, are shared by the single women quoted in this chapter. Reith defines this experience as an arrested flow of time, through which the addict is marginalized from society's temporal order. This temporal experience, according to Reith, is one in which the addicts are no longer involved in the social process of becoming (ibid.).

Many single women reflect upon their lives in a similar fashion. They sense that their future is blocked, the present emptied of meaning. As Shirli, quoted previously (Malachy 2010), exclaims, she misses out on her life while waiting for something to happen. In this manner, her life is devoid of any meanings and she can no longer be involved in the process of becoming.

Reith notes that the common experience of the addict is that life is wasting away, where each day seems like any other day. Accordingly, "nothing is happening"; their daily rhythms are defined by inactivity, repetition, and stasis (Reith 1999). Years of addiction are often termed as lost, barren, and unproductive, and "major" life events, which are normatively viewed as formative, leave no imprint on the addict. Reith beautifully conveys this as a *breakdown in the articulation of time* (emphasis mine) in which "time ceases to be sequential and forward moving; it loses its telos" (ibid., 102).

The findings of Charmaz and Reith reiterate some of the temporal experiences of single women who, as this chapter has shown, sense that their life is on hold and accordingly understand their life as a biographical disruption. The ruptures in the expected gendered life course, during which a woman should marry and have children at a certain time, can also be interpreted as a breakdown in the articulation of time.

The perceptions of single women as immobile subjects can also take insight from Bauman's (1998) critique of globalization. Bauman excoriates the growing gap between mobile and immobile worlds, and the ensuing hierarchy established between consumer-tourists of the first world and the vagabonds of the second. Bauman makes the following claim:

The inhabitants of the first world, i.e., the global elite, live in a perpetual present, going through a succession of episodes hygienically insulated from their past as well as their future. These people are constantly busy and perpetually "short of time" since each moment of time is not extensive. People marooned in the opposite world are crushed under the abundant, redundant and useless time they have nothing to fill with. (ibid., 88)

Bauman elaborates that while the inhabitants of the first world are moving on, going beyond the constraints of time and place, the inhabitants of the second world live in a time in which "nothing ever happens." Although Bauman writes about the different types of temporalities created by the growing global gaps between the rich and the poor, his analysis can shed light on the temporal hierarchy differentiating between mobility and immobility.

This interesting corpus of societal temporal inquiries highlights the extent to which we have become subjects through being embedded in certain kinds of temporality. Employing this analytical perspective, we can discern the various expressions in the breakdown in the articulation of time to temporal identity formation. Thus, the perception of single women as immobile subjects is connected to their location in time and their possibility of becoming subjects and having a future. These possibilities are evidently gendered and heteronormative, depicting a blocked future presently characterized by numbness.

Their experiences of immobility become more perceptible when it seems that others are moving ahead in a linear progressive fashion. The view of single women as immobile subjects also alludes to the hierarchy formed between what can be seen as two temporal discernible positions. If we draw from Bauman's rich formulation, when one is coupled, one can control and transgress time by having the ability to move forward. When one occupies this position, time can be perceived as a resource through which one can live in a dynamic present, and can move ahead towards the future. Couplehood and family life open possibilities to the future. On the other hand, it appears that many single women lose their grip on time and are perceived thus, and as a result are trapped in an extended numbness and immobility.

In the introduction to this chapter, I mentioned Halberstam's observation that queer time highlights "the potentiality of a life unscripted by the convention of family, inheritance and child rearing" (Halberstam 2005, 2). In the works on time mentioned earlier in this chapter and in the course of this book, the presuppositions of what can be considered as normative time are constantly negotiated and challenged. These works emphasize that the hegemonic conventions of time are not absolute but open to change.

Halberstam's work, for example, can prompt us to rethink the dictation of what are perceived as normative rhythms and what is considered to be an a-synchronized temporal experience. As demonstrated throughout this chapter, the experience of single women taking a break is read primarily through what is considered the normative life course dictating a linear, developmental telos. Accordingly, the various accounts unfold the effects of these prescribed temporal templates and tempos. Moreover, single women recognize their a-synchronized social standing yet insist on taking a break. Although some of them obey the temporal heteronormative framework at large (by defining their break as temporary), they nevertheless challenge conventional sequential rhythms and temporal idioms.

Stimulated by this rich critical literature, I propose to envisage singlehood temporality not merely as a non-synchronized timeout or as time on hold, but as a position from which we can pursue alternative articulations to heteronormative rhythms and life schedules. In that way, it can offer a much needed counter-logic to heteronormative temporality. Such a temporality affords long, unlimited breaks and delays, as well as experiences such as being stuck and frozen, which are an inseparable part of our everyday temporal experiences.

Notes

1   See, for example, Karen Stein's (2012) temporal analysis of the vacation.
2   See Lahad (2013, 2014).
3   On the recent debate concerning the offer by multinational corporations Facebook and Apple to cover the costs of egg freezing for their employees, which prompted heated public debate and criticism, see Kuchler and Jacobs (2014).
4   For an excellent discussion of this topic see Hidas (2015).

7

Waiting and queuing

The temporal construct of *waiting* is one of the predominant images associated with single women. The figure of the single woman waiting for coupledom and married life has become deeply embedded in conventional thinking about single women. The "What's new?" genre of questions, the blessing *Bekarov ezlech* ("Soon at yours [wedding]!"), and promises like "By your wedding day you will feel better,"—discussed throughout this book—can be regarded as reflecting and endorsing this temporal imagery. They remind single women of their belated singleness and of their overly extended wait. While waiting for "the right one" at certain stages of one's life is considered romantic and full of hope, at later stages it marks a state of heightened anxiety, stress, and uncertainty.

In this chapter, I seek to trace some of the discursive constructions of waiting and images of waiting single women and, by proxy, problematize these concepts. From this perspective, I look at waiting as both a temporal construct and as an interactional process which sheds light on how power relations, forms of knowledge, and subjectivities are constituted and reified. Moreover, engaging with waiting as a contingent temporal construct also opens up a space to critique the hierarchal relations it creates, and how in turn it creates and maintains power relations.

Hopeful, restless, waiting

Samuel Beckett's (1954) play *Waiting for Godot* famously emphasizes how fundamentally intrinsic waiting is to the human condition. Waiting, adds Giovanni Gasparini (1995), has a wide range of meanings and attributes, and is commonly considered a basic aspect of the human experience. Waiting moves, he observes, from representing a hope and a gratifying experience to a frustration, an illusion, and a form of indefinite distress (ibid., 39).

Indeed, we wait in waiting rooms, we stand in lines, we enroll ourselves on waiting lists. Waiting is a significant part of our social lives and everyday schedules; it is an inherent side-effect of bureaucratic logic and religious beliefs, and is incorporated into a wide variety of social practices. It also plays a central role in our daily social existence and knowledge, as it guides everything from mundane conversation to traffic rules.

Lance Morrow suggests that waiting casts life into a "little dungeon of time" (Morrow 1984, 65). In western capitalist societies, waiting time generally carries pejorative connotations, partly because capitalist society idealizes notions of efficiency and speed, identifying time with money and, thus, waiting with idleness or waste. One often seeks to minimize waiting time or to eliminate it altogether. Accordingly, waiting is associated with bad service and inefficiency. As a result, today significant technological and organizational effort is invested into seeking to reduce waiting time.

Waiting in its romantic formulation is built into our notions of romantic longing, as expressed beautifully in a verse from "The Man I Love," the classic love song written by Ira Gershwin and performed by singers such as Billie Holiday and Ella Fitzgerald:

Someday he'll come along
The man I love
And he'll be big and strong
The man I love
And when he comes my way
I'll do my best to make him stay

The storyline of women waiting to be chosen is set in a long tradition of heterosexual romance (Reynolds 2008, 101). Waiting for him "to come along" and "making him stay" complement the cultural image of a "prince charming" or the "knight in shining armor." Waiting, in this sense, connotes excitement, delight, and fantasy. Even the new modalities of love, in which women exercise more choice in choosing their partners (Swidler 2003), stress the centrality of waiting for "the right one" and looking for one's soul mate.[1]

In her analysis of the love stories section on Match.com, Sharon Mazzarella contends that many of the success stories published on the site "tell the tale of individuals who have been searching their whole lives for the prefect partner, a search which has ended successfully thanks to their experiences on *Match*" (Mazzarella 2007, 25). One can find similar stories on most dating websites, which, as a part of their marketing strategy, highlight tales of single men and women who had waited for years before they found "the right one" on this particular dating website.

Another popular illustration of single women waiting for the one can be found in *The Bachelor*, a successful global television format which was also adapted in Israel. One of the highlights of the show is the Rose Ceremony, during which the single women participants in the program wait anxiously to be selected, and hopefully win the big prize—the heart of the male bachelor. As Andrea McClanahan elaborates, "The Rose Ceremony is the validation point for the women. If a woman receives a rose, she is deemed worthy enough to remain in the game, and she is afforded a sense of well-being or happiness by Alex's [the bachelor] decision" (McClanahan 2007, 267).

A different modality of waiting is depicted in the well-known song "Eleanor Rigby," written by John Lennon and Paul McCartney and performed by the Beatles:

Ah, look at all the lonely people
Ah, look at all the lonely people
Eleanor Rigby picks up the rice in the church where a wedding has been
Lives in a dream
Waits at the window, wearing the face that she keeps in a jar by the door
Who is it for?

For Eleanor Rigby, there is no point in waiting. The rice thrown at the happy couple remains on the floor, a reminder to all those lagging behind. Eleanor Rigby can be understood not only as a song about unrealized romance, but also what could be interpreted as a representation of the overly prolonged wait and eventual lonely death of an "aging spinster."

A comparison of the two songs, "The Man I Love" and "Eleanor Rigby," depicts the existential condition of waiting for the unknown. Each expresses a longing for an unidentified male savior. The subjects of both songs wait for a "necessary" transformation in their life course, yet to occur. However, a comparison of the two songs demonstrates how waiting is dependent upon differing situational contexts and temporal timetables. While the first is considered to be one of the iconic love songs of the twentieth century, portraying an image of romantic longing, the other is noted as a song about loneliness, portraying a desperate, pathetic waiting. The woman represented by the line "someday he'll come along" is still *on time*, while the figure of Eleanor Rigby can be perceived as *off time*.

The two images of single women waiting for men in these two songs—well-known in Israel and worldwide—reflect deeply ingrained representations of singlehood and single women at different stages of their life course. The next extract, published in the Israeli website *Mako*, depicts different waiting modalities. The article is entitled "She is 40 Years Old and She is Still Waiting for the Knight on the White Horse," and depicts the life stories of Ortal Arbeli and Liat Dyan, both almost forty years old:

> Ortal Arbeli was not ready to give up the big dream: a child, a dog, and a house with a fence. It didn't matter to her that she was almost 40 years old. Today she's married with one child and she proves that sometimes it is worth the wait to find "the right one". Liat Dyan is single, she's almost 40 years old and she's still waiting for the one. (Yechimovich 2013)

While one waited for years, the other one is still waiting. In Arbeli's case she waited for her soul mate, and found him just in time. As the text implies, she met someone who was "worth" waiting for. Dyan, on the other hand, is still waiting. Waiting for the one is demarcated by culturally agreed-upon deadlines after which there is no point of waiting any more. This is one reason why what may seem like a dreamy and even sweetly melancholic kind of waiting in the earlier stages of the single woman's life course can evolve into an anxious wait.

In this context, it is important to note that waiting, in common with many of the temporal constructs discussed in this book, entails gender-related differences and age/gender-related role transitions which, in turn, form different temporal regimes and timetables for men and women. The waiting experiences of single women are juxtaposed with widespread images of women as passively waiting while recognizing the pressure of biological clocks and the threat of turning into "old maids." In this figuration of waiting, one's whole existence, social status, and possibilities of belonging come into question. The manifestation of these shifts is represented in the figure of the bridesmaid as a looming presence in the linear timeline of the single woman.

Always a bridesmaid, never the bride

In our much-crazed wedding culture, the bridesmaid is a recognizable social figure perceived as "the next in line" to her marrying friend. She is traditionally a single woman, assigned the role of supporting the bride before and during her big day. The bridesmaid also plays an important role in the secondary wedding market. In the US and many other countries, the bridesmaid's role has evolved into a flourishing market, producing its own commodities like special bridesmaid's matching dresses, shoes, flower arrangements, and jewelry.

Popular culture worldwide is fascinated by this figure, and the bridesmaid's role has become especially popular in some of the most recent Hollywood romantic comedies. One example is Anne Fletcher's box office hit, *27 Dresses* (2008), also screened in Israel, which depicts the story of a serial bridesmaid, with twenty-seven bridesmaid's dresses in her closet already, hoping to exchange the bridesmaid dress for a bride's. Another popular film underpinning this message is Paul Feig's *Bridesmaids* (2011), which garnered much media coverage in Israel. The film featured, as in many romantic comedies of its kind, the unhappy life of the bride's best friend, who is given the role of the chief bridesmaid. As in *27 Dresses*, the film focuses on the miserable life of the bridesmaid while she tries to manage all the pre-wedding events and rituals. Both films end on an optimistic tone, the heroines "hooking up" with eligible bachelors. The premise in this narrative is clear: their role as bridesmaid was temporary and transitional, and their waiting period has come to an end.

The bridesmaid film genre can be seen as part of what Negra terms the *relentless celebration of weddings* in contemporary popular culture targeted at women (Negra 2009, 81).[2] In these scenarios, the bridesmaid is often depicted as the negative mirror-image of the bride. While the bride's life trajectory is celebrated and rewarded, the other's is configured as pathetic and miserable. In these films, the bridesmaid's role is to observe and admire the linear progression she is yet to join. She is represented as an immobile bystander, obliged to publicly account for and justify her status. It seems that the wedding ritual not only secures class and sexual orientation hierarchies, but also produces a clear hierarchy between those "who did it" and those "who are still on their way." In this manner, the images of the bride/bridesmaid create a neat binary opposition, in which one category (the bridesmaid) should transform into another (the bride).

Anglo-American clichés such as "Always a bridesmaid, never the bride" or "Three times a bridesmaid, never the bride" exemplify the social conventions which mark the overly extended presence of the bridesmaid as disruptive to the collective temporal order. In Israeli secular and religious marriage culture, the bridesmaid's role is less structured and visible than in Christian weddings. Nevertheless, the presence of one's best friend, sister, or cousin is a recognized informal social role in Jewish-Israeli weddings and shares many parallels with the social role of the bridesmaid. An abundance of texts in *Ynet*'s Relationship section, portraying the bride's unmarried sister, cousin, or best friend, express the unease, embarrassment, and at times even humiliation associated with attending a wedding when one is still placed in the position of the "yet to be" married sister or friend.

The prototype of the bridesmaid and "the yet to be married" not only epitomizes the waiting experience, but also emphasizes a socio-temporal order in which an imaginary symbolic queue is formed. This temporal scheme is embedded within prevailing expectations of who should be next. In that respect, the "eternal bridesmaid" not only signifies some form of bad timing, but is also unsettling to common temporal norms and codes. To reiterate some of my observations in Chapters 2 and 3, this blessing can be seen as an indication of the so-called orderliness of everyday rhythms, as well as the wish to restore and fix potential irregularities.

In this connection, Moore stresses the importance of defining the collective temporal boundaries and the orderly arrangements for synchronization in our everyday lives (Moore 1963, 52). Indeed, as these clichés imply, playing the role of the bridesmaid for too long disrupts sequential and synchronized temporal orders. The social sanction needs no further elaboration: "Always a bridesmaid, never the bride." The extent to which this form of temporal organization creates and maintains hierarchical relations within the matrix of power relations between single and non-single women cannot be underestimated.

This ritual has achieved much visibility in many popular movies and television sitcoms and movies worldwide. I suggest that the folkloristic ritual of catching the bouquet can therefore signify a social event which conveys a particular temporal map, in Zerubavel's terms, a map which reflects prevailing temporal and age-related expectations (Zerubavel 1985, 14). In order to catch the bouquet, single women are expected to gather together and even playfully compete with one another to maximize their chances of catching the bouquet. Mann (1969) has argued that the queue can be perceived as *a miniature social system* of shared behavioral norms. Pursuant to this analogy, single women's statuses can be measured according to their location in the queue and whether or not they can stand in line at all. By the same token, it is evident how the various clichés and images of the waiting single woman, such as the bridesmaid or the single woman singing and waiting for *The Man I Love*, depict and form such a miniature social system, a symbolic line which conveys clear temporal norms.

Nevertheless, whether the single woman occupies the temporary role of bridesmaid, or is being bid by well-wishers to get married soon, the underlying assumption is that the social practice of standing in line means that one is being taken into consideration. She can compete with others for the attention of a potential husband, and then hopefully enter one of society's key institutions. The bridesmaid is "still in the game," and she has a chance if she is able to catch the bouquet in time. Thus, the perception and experience of waiting is dependent on one's age. As noted earlier, singlehood is constituted differently at twenty-five, thirty-five, or forty-five. However, the particular temporal junctures of time and one's awareness of time become a vital factor. Thus, if at the earlier stages of the single woman's life course waiting can be construed as romantic and a positive tension-builder, as singlehood threatens to turn into a permanent status, waiting can become imbued with dread, fear, and uncertainty.

Bekarov ezlech (Soon at yours [wedding]!)

The age-related interpretation of waiting is also exposed in the well-known Israeli blessing, *Bekarov ezlech!* Addressed to single men and women, it is a blessing usually conveyed by the married to the non-married, most often at weddings, and it expresses the hope that the next wedding will be theirs. The tone of this blessing is commonly confident and affirmative. In the case of single persons, the *Bekarov ezlech* wish does not specify to whom one should be married, but instead refers to the act and the event itself. However, the wish *Bekarov ezlech* is conveyed to single women and men in particular age groups. When a single woman passes what is considered as the normative marriage age, she will probably cease to hear this blessing. It could be suggested that just as there is no bridesmaid above a certain age, in a similar vein one would not wish for a sixty-year-old single woman (for example) to get married soon.

The *Bekarov ezlech* blessing has come to epitomize many of the experiences of single women, in private and in public settings, as evidenced by the next extract:

> As someone who for most of her life was in long-term relationships, I heard this sentence dozens of times. Mostly, I heard it from the bride and the groom, who think that their life choice should match everyone else's lives. Usually the bride hugs me with joy and then yells out drunkenly "*Bekarov ezlech, Bekarov ezlech*" … Nevertheless, only when I moved from the category of a single woman in a relationship to a single woman without one, I realize what a real trauma [the blessing] can be. The experience of hearing *Bekarov ezlech* when you are in a relationship is unpleasant. People have no right to interfere with your personal life. But for the available single woman [the blessing] is a magnifying glass to all that is wrong and unfitting in one's life … It does not matter how successful you are you will always be reminded that in this [getting married] you have failed. (Banosh 2011b)

Noa Banosh's reference to the blessing as a magnifying glass elucidates why some single women view this expected encounter as a nightmare, hell, a trauma. Even when these columns are written in an ironic tone, the common denominator of the texts is that they communicate strong reactions and the experience of feeling trapped. These encounters, in which the blessing is communicated is also a moment in which their singlehood and its discrediting features are made known. In such a context, the options for saving face are limited.

The blessing is a critical point of exposure, pointing a finger at the present and the future. These are moments in which relations of power are enacted between the blessers and the blessed, as they demonstrate the differences in their social status. It is another example in which an order of discourse uncovers relations of power and control.

Consequently, these are moments of heightened self-awareness and reflexivity. These blessings are constant reminders of their long, overly extended wait. A post discussing an advertisement for designed wedding greeting cards, published on an Israeli blog, reveals a common attitude to this blessing:

> Who has not heard this this annoying sentence [*Bekarov ezlech*]? I have been hearing this blessing from the age of 12. At family weddings, there was always some aunt that couldn't resist and decided to worry about my future, as it is never too early for these kinds of anxieties. ... As I married quite young, at 25 I no longer have to endure these sentences but I go out of my mind when I hear it. ... The design team of netanella.com introduces a new project for wedding greetings. If you are invited to a wedding or two in the next months, we cannot help the single men and women from avoiding hearing *Bekarov ezlech*, but we can save you time thinking what kind of greeting cards to write. (Hanick Zikukit Tafus 2011)

This advertisement presents an interesting illustration of the social pressures put on single women from an early age. It discloses how these expectations position girls, adolescents, and women in a perpetual waiting position to enter marriage. The blessing epitomizes the way in which waiting signals an expectation for "things to happen." It is a transitional time fueled with despair: moaning, and expectation. Interestingly, while this scripted interaction is anticipated and well known in advance, its very occurrence is rarely negotiated and contested.

In his well-known anthropological work about waiting time in South Africa, anthropologist Vincent Crapanzano notes that at the mercy of time, the waiting individual is subject to "feelings of powerlessness, helplessness, and vulnerability—infantile feelings—and all the rage that these feelings evoke" (Crapanzano 1985, 44). Although this is quite a different socio-historical context, I argue that these are moments in which one's subjectivity is at stake. This is one of the reasons why it is difficult to resist these powerful social prescriptions, which also determine the possibility of one obtaining a heteronormative future. Waiting, as Crapanzano observes, is being at the mercy of time, a position which makes it hard to resist time and its imperatives. Thus while many of the single women resent the social pressures and hierarchies exposed in these encounters, there are no discursive resources to resist their predicted timeline and heteronormative assumptions. The writers grasp the blessing as a social fact, that "we cannot help the single men and women from avoiding hearing *Bekarov ezlech*."

The fact that this social temporal order cannot be avoided is also represented in Goldy Heart's column:

> Every single man and woman knows that one cannot escape the *Bekarov ezlech* blessing ... nevertheless I want to ask why these aunts, who in certain cases have not seen me since my *Bat Mitzvah*, think they know what I want in my life right now ... To be honest, I don't know if this blessing is intended for me or for the aunts themselves. (Heart 2008)

Waiting to be next, then, is far from a personal endeavor; as mentioned above, the blessing itself labels single woman's wait as a collective waiting project. These social pressures are apparent in both the American and Israeli clichés; *Bekarov ezlech* and "Always a bridesmaid, never the bride" both reflect what is reformulated again and again as a social problem—late singlehood. In the column mentioned above, Noa, a single woman, explains how the happiest day of the bride's life can turn out to be a miserable day in the single woman's life, especially since, as she emphasizes, she dared to turn up by herself:

> So you [referring to the bride and the groom] made me drive to the middle of nowhere and spend a fortune. Please let me suffer quietly on the way to the buffet and do not interfere with my private life ... If there is one thing on which there is a consensus among single women is that we hate weddings. Not our own, the one we have been dreaming about since the age of five but the weddings of other women. The only way in which one could enjoy herself if you arrive with all the right accessories: a dress, high heeled shoes, a wallet and a boyfriend. Now, after clarifying that I do not like weddings ... Let's point at the big elephant sitting under the *Hupa* [canopy] waiting for me to uncover him ... I am referring to those two words that can make every single woman's life a living hell: *Bekarov ezlech*. (Banosh 2011b)

The anxiety expressed here communicates how the expectation of this ceremonial encounter also conveys a loss of control. Being accompanied can provide a shield to this interaction and exposure as a single woman. Thus, her chance to regain control is dependent on finding a male savior. The right man does not only promise marriage, but also a renewed sense of agency and belonging.

In her research about waiting among the mothers of bachelors in Macedonia, Violeta Schubert (2009) writes that waiting for marriage is related to the upward social mobility of both the bachelors and their mothers. For these mothers, the single status of their sons has a significant impact on their daily interactions with other women in the village. Their son's unmarried status causes them to occupy a low status in the village's social hierarchy, and leaves them vulnerable to being provoked by other women in their villages. Waiting is a collective project not only in Macedonia. The status of the unwed son and daughter has a significant impact on parents in Israel as well. In the next account, Bat Chen, a single woman, describes her mother as: "Waiting for the moment when I will tell her that I found him. Without me saying it didn't work; just telling her simply that I found true love" (Bat Chen 2009).

In his well know study *Timetables: Structuring the Passage of Time in Hospital Treatment and other Careers*, Roth (1963) has described how people constantly try to define when things will happen to them and measure their progress according to temporal norms and benchmarks. In the same context, single women's parents are waiting with them; they are constantly measuring their progress according to these benchmarks.

In Noa's column mentioned previously she describes the scrutiny she experiences from her family when attending a relative's wedding:

> Every familial event becomes an opportunity to make you feel bad. Undoubtedly the wedding is the highlight for those. Waiting becomes a nightmare to such an extent. If we single woman had a choice between that and *A Nightmare on Elm Street* we would have chosen Freddy Krueger without any doubt ... Of course the interest in

you does not boil down to the blessing or showing interest which is concerned or pitiful. Your singlehood is thrown in your face, again and again, throughout the wedding. (Banosh 2011b)

As noted earlier, in these familial settings the blessing is perceived as an unwanted interaction, a mechanism through which one's singlehood "is thrown in your face again and again." Returning to Mann's (1969) conceptualization of the queue as a miniature social system of shared behavioral norms, the person who blesses the single woman is evidently not considered to be standing in the same line as her. This encounter implies the tacit hierarchy of a temporal order, thus reinforcing the explicit and implicit boundaries between the person doing the blessing and the person being blessed.

Waiting for the unknown

As noted in earlier chapters, late singlehood is characterized by its main feature, the delay in getting married, a liminal state which has seemingly transgressed and violated its expected temporal boundaries. Against this background, from a certain stage in the single woman's life trajectory, waiting is related to growing personal, familial, and communitarian uncertainty and social anxiety.

The texts analyzed so far depict the difficulties of waiting for the unknown. Waiting becomes a source of suspense and uncertainty also due to its liminal attributes. In his widely quoted study on the ritual process, Turner (1969) argued that the liminal intermediate phase is of fundamental sociological importance. As discussed in Chapter 3, Turner, in drawing on Van Gennep's (1960) theory of the three stages of rites of passage, paid particular attention to the second stage, the liminal phase.[3] Liminality, he emphasized, is a state of being between phases—a transitory position. As such, the individual positioned in the liminal phase is not a member of the group one previously belonged to, nor of the group one will belong to upon the completion of the next rite. In fact, liminal subjects are "neither here nor there; they are betwixt and between the positions assigned and arrayed by law, custom, convention, and ceremony" (Turner 1969, 95).

The widespread images of anxiously waiting single women could be grasped as *liminars*, in Turner's terms (Turner 1969). This understanding corresponds closely with my contention that singlehood is generally framed as a liminal, temporary state; a transitory stage on the way to couplehood and family life. According to prevalent representations, the single, not yet married woman is depicted as waiting, hoping, speculating as to when the liminal period will come to an end. This familiar image is implicit in the following *Ynet* column:

> Where would I meet him? How would it happen? I couldn't let myself believe that I would find him. How could I be optimistic when I had no clue as to the outcome of my search? One of my friends told me that perhaps instead of thinking about how I should think about when … he's out there you don't know exactly where … the only question is when you will meet him and not if you will meet him … it's just a question of time. (Netz 2008)

In the above passage, Tali Netz, a single woman, stresses her liminal and uncertain social position. The liminal stage, as Turner (1969) notes, is characterized by ambiguity and inversion resulting from an anomaly wherein people slip through networks of classification. While marriage is commonly regarded as a charted and planned passage, permanent or prolonged singlehood is often viewed as an emergent, unplanned life trajectory.

Thus, the above quoted paragraph conveys certainty and uncertainty at the same time. The writer predicts that it is just a "question of time" until she finds her one, but she has no knowledge as to where, when, and how. There are no clear temporal references and the exact timing of progress from one temporal position to another is unknown. The experience of waiting becomes ever more intolerable for some of the single women, as the status of being single can change the next day, in a few years, or never at all. As Dazy Bar, another single woman writing for *Ynet*, observes: "I am thirty years old, six years past the age I was supposed to be married, and there is no potential groom on the horizon" (Bar 2009).

This position can also be perceived, then, as a body of clues, constructing the norms of collective timetables (Roth 1963). Indeed, at some vague and unstructured point in time, singlehood shifts from being a socially legitimate temporary phase to what can be characterized as a biographical and social disruption (Bury 1982). In other words, lifelong singlehood marks an unexpected disruption of a seemingly normative liminal state which has unexpectedly become permanent. The texts analyzed here demonstrate how waiting is interwoven with the wish and the social pressures to leave the liminal territory of uncertainty and vagueness, and to enter a non-liminal state.

Prolonged liminality and uncertainty

Scholars writing about waiting emphasize how it is dependent on the possibilities of mastering the unknown. Javier Auyero, for example, has documented the relationship between waiting and uncertainty experienced by welfare recipients obliged to endure endless arbitrary postponements, bureaucratic mistakes, and changing state requirements: "In the recursive interactions with the state, poor people learn that they have to remain temporarily neglected, unattended to, or postponed" (Auyero 2010, 857).

This is reminiscent of Schwartz's (1975) observations with respect to what he discerns as the relation between waiting, punishment, and power relations. For Schwartz, punitive sanctioning through the imposition of waiting is met in its most extreme form when people are not only kept waiting, but are ignorant as to how long they must wait. The person then finds himself in an "interactional precarious state wherein he might confront, recognize and flounder in his own vulnerability or unworthiness" (ibid., 38). Thus, summarizes Schwartz, waiting is the crossroads not only between past and future, but also between certainty and uncertainty.

Schwartz's evaluations can also be applied to the *Bekarov ezlech* (Soon at yours [wedding]!) blessing referred to in the next extract:

> Do these people have a special calendar from which they know the specific date that
> Goldy Heart will marry? Just tell me; I promise not to get mad if they do. It seems to me

that these kinds of calendars and crystal balls only exist in Harry Potter films, and so these kinds of blessings are particularly annoying. They attempt to promise something which is beyond the control of the person who is blessing me. Can you promise me a specific date? If so, then fine; promise. *Bekarov ezlech* is simply not good enough. (Heart 2008)

The blessing lays emphasis on the manner in which our social life is constantly organized and regulated by temporal schedules and temporal boundaries. In this case, the career timetable of the single woman is prescribed in advance, and social injunctions therefore spur her on to move forward in a predefined and recognized linear trajectory, in which marriage is the ultimate goal. Nonetheless, the incorporation or reincorporation of the single woman into society marked by finding one's soul-mate and building a family may or may not happen. Indeed, prolonged singlehood is regularly represented as a period of growing uncertainty and instability. These clichés therefore provide important signposts and, in this case, structure, and bestow meaning upon the passage of time.

By the same token, Crapanzano has observed that waiting implies a particular orientation in time, directed toward the future; nonetheless, it is a constricted orientation that closes in on the present:

> In waiting, the present is always secondary to the future. It is held in expectation. It is filled with suspense. It is a sort of a holding action ... in waiting the present loses its focus in the now. The world in its immediacy slips away, it is derealized. It is without élan, vitality, creative force. It is numb, muted, dead. Its only meaning lies in the future—in the arrival or the non-arrival of the object of waiting. (Crapanzano 1985, 44)

Crapanzano notes that in English one cannot distinguish between waiting for something concrete and waiting for anything to happen: "in waiting for something, anything to happen, the object of the intentional act of waiting, like the object of anxiety, is not given" (ibid., 46). In this symbolic line, the single woman does not know exactly if and when she will reach its end. It is unclear to the single woman and to the observer whether or not the queue can be beaten and whether there is any potential for queue-jumping, queue-drifting, or leaving the queue altogether. Therefore, a corresponding social division is fabricated for the waiting single women by the non-waiting, not-single women, who presumably do not have to stand in line anymore. A particular temporal framework is constantly formed and reformed, embedded within explicit and implicit cultural beliefs about societal and temporal norms and expectations.

Given the above analysis, I suggest that singlehood as a prolonged or permanent liminal status differs from other liminal phases, due to a fundamental vagueness as to its end point. Think, for example, of a Ph.D. candidate submitting a request for a scholarship. One generally knows when one can expect an answer and can plan ahead accordingly. On the other hand, the temporal location of the single woman is uncertain; she cannot determine how soon she will arrive at the end of her wait. As opposed to Turner's (1969) conceptualization of liminality, in which one stands between two clearly defined stages of separation and re-aggregation, the exact point of re-aggregation in this case remains largely unknown, resulting in a particularly stressful waiting experience.

The production of the waiting subject

Feminist scholarship has long demonstrated how and why women are defined in relation to men and in terms of their relationships to men. Girls and young women are perceived as daughters, wives, and mothers. From this dominant perspective, single women are defined in terms of their lack of relationship to men. The study of the temporal concept of waiting adds another layer to this analysis. Women are socialized, from a very early age to wait for the right man, as Billie Holliday and many others have articulated: "Someday he'll come along, the man I love."

One of the etymological definitions of waiting describes it as a state of alertness and of having a heightened sense for changes. In German, the etymological meaning of the word is to watch and to guard; in English, to be awake. This kind of alertness and being on guard is both personal and collective with regards to single women of marriageable age. Moreover, this alertness finds expression is the constant social surveillance that they are subjected to. Single women are forever being questioned: "So what's new?" "Are you seeing anyone?" "What are you waiting for?" All of these familiar utterances, I suggest, also reflect and enhance the hierarchical relations embedded within the idea of the waiting single woman. My argument is these assumptions give expression to heteronormative logic and produce power relations supported by a disciplinary temporal regime that differentiates between the waiting single woman and the non-waiting, non-single woman.

Schwartz argues that the distribution of waiting time coincides with the distribution of power (Schwartz 1975, 5). In this respect, waiting mirrors temporal power relations: there are those who wait and those who are waited for. To be kept waiting is a social assertion that one's time and social worth is less valuable (Schwartz 1975). As already noted, single women above a certain age symbolize a disruption of the sequential rhythm of our social lives. As Moore elaborates, "The sequential ordering of activities provides a priority schedule in the strict sense, which may reflect priorities in the loose sense of relative values" (Moore 1963, 48). One effect of this feature is the marginalization and subordination of single women. In this respect, the image of the waiting single woman reflects such a rigid form of sequential ordering, representing, and producing temporal orders. These almost unnoticed miniature systems lie at the heart of the socio-temporal discursive formations and temporal monitoring of single women.

As my analysis in the next chapter will show, public events as weddings, Valentine's Day, or New Year's Eve celebrations locate single women in a particularly vulnerable position. When they receive the blessing, to be married soon, their waiting is assumed and repeated again and again. From an Althusserian (1971) perspective, the *Bekarov ezlech* blessing can be seen as a strong moment of interpellation. Weddings and familial gatherings can be seen as spaces through which ideologies turn individuals into subjects. Being blessed resembles Althusser's well-known analysis of the call of a police officer towards an individual: "Hey, you there!" or "Papers please," which requires one to recognize oneself as a subject. This is one of the reasons which could explain why some single women dread this interaction and take it so seriously. This is a critical moment, in which they are interpellated as waiting subjects.

From this perspective, waiting, as Bourdieu (2000), Schwartz (1975), and many others have adeptly identified, is an exercise of power that plays a significant role in the way subjects become compliant subjects in everyday life.[4] Bourdieu observes that waiting is one way of acutely experiencing power—a form of submission, and in that respect waiting is one of the prominent effects of distributions of power. From these works, we learn that waiting is a form of submission, and no wonder that is assigned to the weak, the poor, and the subaltern (Jeffrey 2010; Vitus 2010). This chapter has dealt with a prevalent discursive representation, according to which the single woman is perceived as a bystander, a candidate, and a passive daydreamer, waiting for the unknown. Waiting becomes *a mode of being*.

The various social encounters, such as the lineup of the bridesmaids or being blessed with *Bekarov ezlech*, form, I argue, a *symbolic heteronormative queue* enmeshed with disciplinary power relations and forms of control. From this standpoint, when one hopes for single women to soon be married, this expectancy forms part of a normative injunction emphasizing a linear social order and the way it positions single women within collective timetables. This form of horizontal and vertical lineup is also represented in the symbolic figure of the bridesmaid and accentuated during social events such as the catching the bouquet ritual. It is astonishing to realize that little has changed in the last fifty years, with regards to perceptions of the subject position of single women as women in waiting. This, I suggest, demonstrates the strength and persistence of the heteronormative temporal ideologies, which promote an ideal image of women as good wives and mothers. In the next chapter, I set out to further explore the temporal interactional elements of being and appearing alone in public. This line of analysis will I hope shed more light on the power hierarchies ordained by heteronormative temporality.

Notes

1   Trimberger (2005, 2) observes that many women wait to find their soul mate. According to this prevailing ideal, the soul mate is someone with whom one can combine love, fidelity, emotional intimacy, and togetherness.

2   For an excellent analysis of weddings in popular culture, see Ingraham (1999).

3   According to Van Gennep (1960), the first stage—the pre-liminal—is a state of separation, of detachment from societal structure or from relatively stable cultural conditions. The second—the liminal—is the interstitial phase or the margin, and the third—the post liminal—entails the reentering of the social structure.

4   See also, Auyero (2010); Hage (2009).

8

Time work: keeping up appearances

Over the years that I have researched Israeli internet portals, I have detected a repetitive, periodical movement. As holidays like *Rosh Hashana* (Jewish New Year's Eve) and *Passover*, or widely commemorated romantic celebrations like Valentine's Day approach, Israeli websites begin to publish a range of columns, written by and about single women, discussing their fears of being—and appearing to be—on their own over the holidays. This phenomenon is not unique to Israeli society, of course. One can easily find any number of similar posts on American or British websites and portals, recounting the loneliness of the single woman during the holiday season, or routine social embarrassments such as dining alone in a restaurant or going out alone to a bar.

Many dating and relationship experts publish tips advising readers how to cope with the holiday period: facing one's immediate family with confidence for instance, what I would describe as *keeping up appearances* as a single person. Some columns advise their habitués how not to fall prey to the self-pity and angst that can accompany spending Christmas or New Year's Eve alone, while other writers suggest witty responses to impertinent questions from family members like "When are you going to settle down and give me some grandchildren?"

The pressure is both explicit and implicit, verbal and non-verbal. Single women above a certain age report the surprised or pitiful gazes directed at them during family gatherings, and often complain that they are constantly forced to account for their enduring single status. My analysis of web columns over the last eight years shows that appearing alone in public in couple- and family-oriented societies leads one to heightened reflexivity and, when possible, serves as an impetus towards the careful management of one's social appearance. At particular times and in particular settings, single women are made particularly aware of their required performance, and of the temporal norms that impede their appearance in public.

This chapter reflects upon this dynamic from an analytical perspective, one which takes into account the temporal interactional elements of being and appearing alone not merely within familial settings but also in other public settings like bars, café, New Year's Eve celebrations, and work-related functions. As I will show, single women are particularly aware of both the temporal rules and of the ensuing assumptions that these rules thrust into their everyday lives. I argue that the temporal elements of social situations such as New Year's Eve, Valentine's Day, the weekend and going out for dinner have a significant impact on the visibility of single women, and affect their ability to orient and assert control of their agency in public settings.

In general, my exploration of this dynamic will tend to a Goffmanian analysis, in particular drawing from his conceptualizations about the *interactional order in public settings*. Such an approach offers a means through which we can understand how perceptions of social time produce both the societal freedoms and societal restraints that guide—and restrain—the presentation of the self in public. The presentation of the female single self in public is, as I shall demonstrate, very much dependent on the conventions of social time. In this sense, this chapter also aims to make a significant contribution to symbolic interactionist literature, by exploring the temporal elements of the interaction order.

Beyond this, the situational and interactional analysis presented in this chapter emphasizes the links between *temporality, relationship status*, and one's *interactional unit*. My understanding of temporality rests upon an examination of interactional dimensions, and vice versa. The sociological understanding—widely accepted—that during everyday interactions social actors attempt to control the information others have about them should be re-evaluated, I argue, by taking into consideration the temporal components of these interactions.

My analysis shows that at certain times of the day, the week, the month, and the year, familial and heteronormative codes are particularly reinforced. This is one reason why many single women report an increased intensity in the regulatory gaze towards them at these times. By implication, these are times when single women become particularly self-reflective and aware of temporal social protocols.

This chapter will explore the temporal regularities of everyday life from a different perspective. It will take into account the temporal interactional elements of being and appearing alone at particular times (such as night and day, the week, and the weekend). Thus I argue that the temporal elements of social situations, such as New Year's Eve or dining alone, have an important bearing on single women's impression management.

"The holidays are difficult for singles"

In 2010 the Israeli Channel 2 news reported that some Chinese single men and women had found an original and expensive solution to cope with their parents' criticism: rent a date for the Chinese New Year's eve (Channel 2 News 2010). Following this story I discovered that two years later the *China Daily* published that Taobao.com—a major e-commerce website in China—offered a "rent a date" service, providing a companion for single people to take to their parents' homes on Chinese New Year's Eve (China Daily 2012). Covered by news agencies around the world, the story's main emphasis was the fact that the service was in demand. In an interview with the *Guardian*'s Beijing correspondent, a twenty-six-year-old single woman explained:

I was not looking for some perfect guy to marry. Just someone tall—my parents like tall guys a lot—honest and not too talkative, so he doesn't say something wrong ... My parents want me to get married by 30 ... Bringing a "boyfriend" back home simply means I get less hassle from relatives and my parents will stop worrying about my romantic life. (Branigan 2012)

From a Goffmanian perspective, the fake boyfriend plan can be interpreted as a strategy for interaction, which enables single people both to avoid familial criticism and to save face (Goffman 1967). The interviewee understood that New Year's Eve always created a precarious experience, and made her decision "in the light [of] one's thoughts about the others' thoughts about oneself" (Goffman 1969, 101). In this respect, she can be viewed as a strategic actor aspiring to exercise control over the impression management they convey to others. Hiring a boyfriend enables her to create the right kind of image for her audience—in this case her parents. Being single demands a carefully planned performance.

The wide reportage of this new commodity strikes a chord. Amongst other things, it reinforces another claim made by Goffman, that for single people some interactions are precarious events, during which they struggle to save face and maintain their dignity. I have yet to find such initiatives as the Chinese rent-a-date service in Israel, yet one can find similar websites, like Dates4Hire in the USA, offering companionship services and escorts for a range of social events including weddings, proms or work and family related functions.

On their website, they state:

Dates4Hire was created with one purpose in mind and that is to provide people the ability to hire a platonic date on demand. Living in a fast paced world and dedicating most of your time to your job and career doesn't always leave you very much time to pursue a romantic relationship.

   Nevertheless, just because you're single doesn't mean you should have to attend social and family functions on your own. Hiring a date from our site is not only as easy as clicking a mouse, but also gives you the ability to choose someone that is compatible to your event. The other major benefit of hiring a date is that they are there to provide a service to you and make sure that your night is a success no matter what the situation. (Dates4Hire 2014)

Both services exemplify the Goffmanian principle that our interactions require performance, in this case a boyfriend or a date for hire is the desired prop. It is no longer a problem if one is single, as Dates4Hire stress that they can provide the necessary props and ensure a successful performance. This theme is reinforced in many of the columns written by single Israeli women. In a column published the day before Valentine's Day, "For Those in Love, Every Day is a Day of Love—But What About Me?," Lalli Blue (a pseudonym), a single woman, describes the experience thus:

A few times a year, let's say New Year's Eve, Valentine's Day, your ex-boyfriend's birthday and your younger cousin's wedding—she is at least one year younger than you—you, as a typical single woman, have to go through the ultimate singlehood test. (Blue 2007)

The holidays, according to the writer, are "the ultimate test" for single women. The derivative question is, why does she refer to this as a test and—more importantly—what is at stake? In other words: why and how is she tested? Why do these special days and rituals pose particular challenges for single women? What is the essence, and the purpose of the test?

As noted above, as the holiday seasons approach, various experts proffer advice about coping with the stress and depression that is part and parcel of the season. Odeta, a well-known Israeli columnist, elucidates: "The holidays are a special period when the single population are most aware of their single status, as they don't have a partner to take home [to their family]" (Odeta 2004). A different column, by Adi Kimchi, a dating advisor writing for *Ynet*, bears the title: "Passover and You Are Alone. How to Cope with Your Family?" The column's opening paragraph runs thus:

> Why, on an evening which is supposed to be harmonious and familial, do you get the feeling you are being criticized more harshly than ever, the fact that one is alone multiplies itself, and a feeling that you are being judged continues throughout the dinner like horseradish burning your nostrils? This is a column about the Jewish genome, with five tips on how to cope with this situation. (Kimchi 2014)

Yael Doron and Gili Bar, also relationship advisors and columnists with *Ynet*, appropriate a client's thoughts about Valentine's Day:

> It's Valentine's Day, and once again the only telephone call I will get will be from my mother asking me the same question: "Well, what's going on? Is there anything new?" I'm so depressed ... All the guys whom I've met so far have either broken my heart or only wanted sex. Actually, when I think about it, I'm better off on my own. Could it be that I don't have this couplehood gene? Perhaps people like me aren't supposed to be in relationships? Perhaps I'll have a child on my own. (Doron and Bar 2007)

In the extracts above, the holiday season and Valentine's Day prompt acute self-awareness. Elizabeth Sharp and Lawrence Ganong (2011), who interviewed single American women about their experiences of singlehood, describe these occurrences as *encountering triggers*. The women whom they interviewed perceived couple-oriented holidays like New Year's Eve and Valentine's Day, and family-oriented holidays like Thanksgiving and Christmas, as the triggers that reminded them of their single status.

The point I wish to emphasize is that one's self-awareness is connected to one's temporal awareness, and to the particular cultural scripts dictated by conventions of social time. As Eviatar Zerubavel (1981) clarifies, the temporal regularities of our everyday lives are among the major background expectancies that shape the basis of the "normalcy" of our social environment. Taking this into consideration, the holiday seasons are often perceived as times when heteronormative familial ideologies take center stage. These ideologies promote the family values that emphasize the primacy of the familial unit, familial togetherness, and family bonding. These are the times that not only is family *done* (Morgan 1996) but also has to be *displayed* (Finch 2007).

A time for display

In recent years, sociologists like David Morgan (1996, 2011) and Janet Finch (2007) have created new analytical tools for the understanding of the lived experience of family life—tools that stress that the family is a constructed quality of human interaction, defined through its activities. Finch argues that displaying families is "the processes by which individuals and groups of individuals, convey to each other and to relevant audiences that certain of their actions do constitute 'doing family things' and thereby confirm that these relationships are 'family' relationships" (Finch 2007, 67). To this, she adds that an important message conveyed to external audiences is that "This is my family, and it works." My contention in this context is that temporal regularities and their background expectancies are a significant component of this message.

Hence, when considering the extracts above together with Finch's and Morgan's observations, I argue that time plays a crucial factor for both doing and displaying families. The holidays, which often include family meals with extended family members, contain many ritualized aspects which are intensified by a scrutinizing gaze directed towards those who do not conform with its normative structures and temporal rhythms.

Single women do not do family, and neither can they put one on display. In his evaluation of Finch's theory, Heaphy (2011) claims that displaying families cannot be disentangled from the normative ideals of a white, middle-class, nuclear family. What counts as a good and convincing display depends on one's subscription to familial norms. The single woman's presence reveals these normative elements, placing emphasis on the normative parameters that underpin what would be considered as a successful display. In these settings, the single woman stands as an uneven number, herein defying the social protocols of the normative and required components of family time and family togetherness.

In a similar way, Valentine's Day and New Year's Eve parties can be considered as times during which coupledom is done and displayed. In a culture in which, as Shelly Budgeon (2008) points out, heterosexual couples occupy a privileged position, events ranging from family meals during holidays to Valentine's Day, are times where this position is recognized and receives its social symbolic reward.

From this viewpoint, for some single women these are the periods when the fact that they lack the required and privileged coupled status is accentuated. It is noteworthy that the increasing commodification of the holiday seasons in public culture contributes in many ways to the increasing visibility of couplehood and family life.

Yael Doron, a dating advisor, tells the story of a thirty-five-year-old single woman, which reveals the difficulties inherent in spending the holiday season with her family:

> I'm fed up. Everybody thinks that I should help, from early in the morning, with preparations for the festive meal. The reason is that all my siblings arrive with their life partners or children. I'm fed up that everybody assumes that I should wash the dishes and tidy up afterwards. The reason for this dynamic is that everybody assumes that I am not in a rush to go anywhere. I'm fed up with being blessed and prayed over again and again [to

get married] ... I'm fed up of being thirty five years old and feeling old and desperate. I'm just fed up! (Doron 2010)

The holiday season is depicted here as a time of crisis. This account manifests a division of labor, organized according to one's relationship status. In family encounters, coupled family members enjoy certain privileges. As the above quoted single woman points out, the unjust allocation of domestic chores is linked to her status as a single woman, unable to enjoy the privileges granted to her brothers and sisters.

Ann Byrne (2003) describes a similar paradigm, as experienced by single women living in Ireland. Some of the single women interviewed by Byrne reported that around their families of origin, they felt like second-class citizens, invisible and less important than their siblings (ibid., 454).

The account quoted above emphasizes that Israeli single women are not only denied the privileges granted to their married siblings, but have their time devalued too. In the accounts above, single women's time is considered as less valuable, thanks to the presumption that due to their single status, they have "no life of their own." One might assume that this unjust division of work is also related to different gender-based expectations; even so, her single status is the prism which she lays emphasis upon, and which coheres with similar accounts from single women.

The advisor describes this single woman as:

> Sitting on the sofa and crying. She is so beautiful and successful. She has a car, owns her own apartment, has a good job and has a promising future yet she is so miserable and desperate. The holidays are always difficult but from one year to another it seems that her capacity to cope with the holiday season decreases. For her coping with the holiday period becomes harder and harder. (Doron 2010)

I propose that one's self-perception—the experiencing of increasing social visibility and invisibility alongside the devaluation that accompanies this—cannot be understood without paying attention to the social meanings of time. Another point at issue here is the importance of time units, and the way that we, as social actors, differentiate between them. Such an analysis leads us to Emile Durkheim (2008) and his consideration of the separation between religious and profane life. As Durkheim argues:

> The religious and the profane life cannot coexist in the same unit of time. It is necessary to assign determined days or periods to the first from which all profane occupations are excluded. Thus feast days are born. There is no religion, and consequently, no society which has not known and practised this division of time into two distinct parts, alternating with one another according to a law varying with the peoples and civilizations; as we have already pointed out, it was probably the necessity of this alteration which led men to introduce into the continuity and homogeneity of duration, certain distinctions and differentiations which it does not naturally have. (ibid., 308)

Durkheim's distinction between profane and sacred time is of real significance when one attempts to understand how the unaccompanied presence of single women is interpreted in public life. The division of time, to varied time units distinguishing between the everyday and the sacred, is embedded in temporal protocols and temporal

norms. When considering these forms of separation, it is also important to acknowledge what Goffman perceives as the basic units of public life, the *single* and the *with*
(Goffman 2010). These interactional units, as I will now argue, play a crucial role in
single women's ability to plan and master their social performance.

The temporality of participation units

Goffman adds another important layer to the analysis of social time and singlehood.
According to Goffman, our routine participation in public life is conducted through
the distinction between what he grasps as fundamental units of public life: the *single*
and the *with*.

> Individuals navigate streets and shops and attend social occasions … either in a "single"
> or in a "with." These are interactional units, not social-structural ones. They pertain
> entirely to the management of co-presence. I take them to be fundamental units of
> public life.
>   A single is a party of one, a person who has come alone, a person by "himself," even
> though there may be other individuals near him … A with is a party of more than one
> whose members are perceived to be "together." (Goffman 2010, 19)

The different accounts analyzed above show that the temporal dimension of participation units impact upon one's social interaction. The demarcation of time into
ordinary and extraordinary time has much bearing on the visibility of the participation
units to which single women belong, at different times and in different social settings.
  Throughout the texts, single women reveal their hesitations about the obstacles
attendant to being alone in public. Commonly, they are perceived as alone even when
there are other individuals, like family members, friends or acquaintances, near them.
This dynamic demonstrates the somewhat automatic identification between the category of singlehood and that of being on one's own.
  Kinneret Tal-Meir, a dating advisor writing on the *Ynet* portal, describes her encounter with Sharon a few days before *Rosh Hashanah* (The Jewish New Year's Eve):

> My first meeting this morning was with Sharon, just a few days before Rosh Hashanah
> … Sharon is a 36-year-old single woman who is terrified of the holidays. Her younger
> brother is married and expecting a child; her younger sister, about to finish high school,
> has a boyfriend. She is the only one alone. In the present situation, as she explained, she
> has no choice but to be lonely and miserable throughout the holidays. Her married
> friends will be estranged to her, and she will be obliged to watch her happy siblings in
> her parents' house and feel that she is the least fortunate. (Tal-Meir 2013)

She comments further about the effects of the holiday season on the single
population:

> During the holidays, the difficulties of being single are intensified. Everybody arranges
> themselves in couple and family units, and only you are by yourself. The pitiful glances
> will reach their peak on Rosh Hashanah itself. Even the aunt that got married late [but
> not *that* late] will give you that look that will make you feel uncomfortable. Your parents

will do their best to avoid looking at you, yet your mother will talk with your aunts and look at you with sadness in her eyes. Eventually, she won't be able to hold herself back, and she'll ask "Well, you still haven't found someone good enough for you? What's going on with you?" And as a result, you won't know where to bury yourself, whether you want to be in a relationship or not. The New Year holiday season can really be a burden. What should you do? (ibid.)

Another example of this dynamic is the matter of attending New Year's Eve parties as a single woman. Thelma and Louise (the pseudonym of two single women columnists writing for *Ynet*) write about this matter:

The leaves are falling, the temperature is dropping … the typical single woman takes a break from her existential reflections and invests most of her efforts in thinking and preparing towards one evening. (Thelma and Louise 2006b)

In another column, dating advisor Adi Kimchi quotes the experiences of one of her single female clients:

Tonight, everyone will celebrate the end of 2013 and the beginning of the New Year. On these occasions, I always get the feeling that everyone is doing something exciting. It's crowded everywhere, everyone is going out, everyone is partying. You can hear laughter from inside the houses, and the kiss at midnight, oh this kiss, creating an illusion about what the next year will look like … This can be an extremely difficult, even depressing time for those who are on their own. For single women and men, this night is a reminder that they are on their own. That, again it didn't happen. Love has not arrived.

Then just one moment before you begin to go through your telephone book or check the invitations you have received through Facebook … How about going out on a date? Yes! A date on New Year's Eve! Perhaps you'll be lucky and get a kiss at midnight or even more than that: you'll be able to tell your grandchildren that your first date took place on the cusp of 2013 and 2014. How can you do it? Continue reading. (Kimchi 2013)

In her column, mentioned above, Lalli frames this dynamic within the context of an exclusive party: "So it's true that you like to party … but you really don't like closed parties which you weren't invited to. Especially those where you have no one to dance with" (Blue 2007).

Between them, the quotes above create a rich and textured description of the social interaction between single women and their environment during the holiday season. These descriptions are ubiquitous in the blogosphere, and are also present in studies about the lives of single women elsewhere (Byrne 2003; Sharp and Ganong 2011). For example, Sharp and Ganong, who interviewed single American women, perceive these encounters as affecting their experiences of displacement from their families of origin.

As various columnists note, these difficulties become even more acute during the holidays. The observation that the holidays are a difficult time for single women is commonplace, but yet I think it demands further problematization: What changes during the holidays, or on Valentine's Day? As Lalli Blue explains, "everyone looks more coupled" (Blue 2007); the consequence is that the solitary presence of the single woman becomes more visible. Moreover, the difficulties in escaping the familial gaze, and in being subjected to public scrutiny during the holiday season, emphasizes the fact that the single woman's performance does not fit the ideals of a couple-oriented culture. The single woman's solitary presence becomes hyper-visible, and turns into a matter of public concern, prompting intrusive questions.

Employing a Goffmanian perspective, we can argue that the single woman fails to appear in the adequate interactional unit, her solitary appearance disrupting the couple- or familial-based interactional order. During these sacred times so connected with familial values, single women are required to present themselves accordingly—in this case, being members of the *with* participation unit.

In Chapter 7, I analyzed Noa Banosh's column (2011b), in which she opined that one reason why single women hate weddings is because they are expected to arrive with the right accessories—a boyfriend being very much obligatory. The moment of arrival is critical, as it reveals the participation unit to which the single woman belongs to. As in many social encounters, the interactional codes of participation in weddings are couple-oriented, and require—as Noa states—arrival with the correct accessories, in this case belonging to a couple interactional unit.

This is also one of the possible reasons why Noa cites attending a wedding as a single woman as "a magnifying glass that enlarges and reflects all that is wrong and flawed in her life" (Banosh 2011b). Or, as Lalli terms it, "the single woman's ultimate test" (Blue 2007). It would seem that appearing as a party of one leads to a heightened reflexivity and social visibility.

Evenings and weekends

Next, I propose to shed light on the links between participation units and social time from a different perspective. The significance of participation units also brings to the forefront the transition from the weekdays to the weekends, and from daytime to nighttime. Shirli Farkash, a single woman and a regular commentator on the *Ynet* portal, writes about how and why so many single people are afraid of the weekends:

> I know many men and women who become anxious as the weekend approaches. These are people who work very hard: they fight corruption, stand on a stage, they go to court and present their arguments before tough judges. But at the weekend, they fall apart. The fear of being alone for 24 hours kills them, seeps into their soul from Thursday afternoon.[1] (Farkash 2007)

In his study exploring the invention of seven-day week, Zerubavel observed that the concept of a week is "an artificial rhythm, created by human beings totally independently of any natural periodicity" (Zerubavel 1985, 4). Building on Durkheim's observation, Zerubavel regards the week as a cycle of periodic alternations which distinguish between ordinary and extraordinary days. If there was no contrast between them, he observes, there would no rhythm to the week. Shirli—whose column is quoted above—writes about how this contrast affects the experiences of single people. The temporal perspective, she suggests, is crucial to understanding this phenomenon, as these days are identified as the time to be spent with one's family or life partner. In this instance, Shirli notes, "The fear of being alone for 24 hours kills them": alternatively, we can say that a weekend alone is grasped as "time to be killed."

The demarcation between weekdays and the weekend, and between the everyday and holidays, is illustrated, for example, in a well-known Israeli song *Shabatot Vehagim* (Weekends and Holidays). This song, by Yehudit Ravitz—one of the most popular female Israeli singers of the last three decades—reveals the longings of a single woman to be with her married lover during the holidays and on the Sabbath. The woman depicts herself as standing alone on her roof at these moments, imagining her lover with his wife and kids. The implication is clear: these are the times, without question, when one should be with one's family.

The situational and relational aspect of these temporal dynamics and subtle boundaries is brought to the fore in a question posed in another *Ynet* column: "Dear single women and men, have you ever gone out to a bar by yourselves? I haven't" (The Naked Truth 2008). This question raised by The Naked Truth—the pseudonym of a regular *Ynet* columnist—describes the general discomfort experienced by many single women when they want to go out by themselves. This discomfort is rarely problematized. In fact, such a discomfort is prescribed by couple-oriented prescriptions. Moreover, this temporal order achieves a high degree of orderliness and normative expectations. As Goffman elucidates, the interaction order is predicated on a large base of shared cognitive presuppositions, if not normative ones, and self-sustained restraints (Goffman 1983, 5). These normative presuppositions and self-sustained restraints are evident in many of the texts informing the manner in which our social interaction becomes a collective and performative achievement. The writer of the column answers her own question with an experience-based explanation:

> Usually, I go out [to a bar] with a male friend, with a female friend or a group of friends … I don't have a problem sitting by myself in a café with a book, and I very much enjoy being by myself in my apartment and enjoying the silence. But there is something that intimidates me about going out to a bar by myself. I'm afraid that if I go to a bar, it will be seen as a declaration that I'm desperate—alternatively, it may seem that I have failed in persuading even one person to join me. This stands in contrast to my favorite cafés, where I can hide behind the pages of the book and pretend that my loneliness is a result and outcome of my own choice. (The Naked Truth 2008)

This illustration provides a vivid example of how single women become attuned to the cultural scripts dictating the spatial and temporal dimensions of the participation unit one belongs to. Going to a bar in the evening is an activity in which the required participation unit is of the *with*, while being by yourself in a café as a single woman is a valid option.

Thus, going out as a *single* at times identified as *with* time defies temporal norms and regulations. These norms clearly define—and thus to a certain extent, determine— one's capacity for successful self-presentation and options of impression management (Goffman 1959). Appearing on one's own at a bar conveys—as The Naked Truth states—a message of desperation, signaling that you have not succeeded in persuading anyone to join you. Hence, the mere appearance as a single woman at the wrong time and place could be interpreted as an indication of a failed performance, putting herself at risk of losing face.

In her study on Norwegian solo travelers, Bente Heimtun (2012) distinguishes between three social identities available to them during their travels: the friend, the loner, and the independent traveler. Heimtum defines the loner position as one in which a woman travels alone and predominantly feels socially excluded, as opposed to the independent traveler position, in which one feels autonomous and empowered. In analyzing the loner position, Heimtun further notes that touristic spaces can turn out to be familial and heteronormative settings, and in that sense exclude single women and mark them out as "others." As one of the women interviewed by Heimtun declared, "Dinner alone is the worst … you feel left out" (Heimtun 2010, 138). Another woman described the discomfort caused by being stared at while eating alone: "God, they are staring at me. Because they see that I am here alone" (Heimtun 2012, 9).

However, Heimtun stresses that the social identity of the independent traveler is about the enjoyable solo holiday, which underlines the opportunities and freedoms inherent in being single. For them, solo travel presents the opportunity to explore new territories, to meet new people, and to exercise their independence and freedom. Similar experiences can be found in a plethora of single female blogs published in recent years, in which women detail the myriad benefits of traveling and having time on their own.

Yet the Israeli accounts I have found mainly demonstrate experiences of social exclusion and increased visibility during evenings, weekends, and holydays. The writer The Naked Truth notes that even though she often goes to bars with friends, going on her own is not considered by her to be an option. Thus, the question addressed to single women "Have you ever gone to a bar by yourself?" is also related to the risk of losing control of one's ability to manage their self-presentation in relation to others. These public encounters, whether in a bar or in a café, are temporally patterned practices for which the single woman needs to plan carefully.

According to Goffman: "When an individual enters the presence of others, they commonly seek to acquire information about him or to bring into play information about him already possessed" (Goffman 1959, 2). Time plays a significant role in this kind of social encounter. The single woman has to play and plan her performance in this ritualized setting. The point I wish to emphasize here is that entering a café in the morning or the afternoon does not require such careful strategizing. Thus, as we can see, social normativity is socially produced through conventions of time. According to this interpretation, if the single woman decides to sit with a book in a café, she can minimize the risk of embarrassment and possibly have more control of the impressions formed of her by others (Goffman 1959). Additionally, she is fully aware that by going out to a bar on her own she might not be able to control others' responsive treatment of her, and consequently risks being embarrassed and ashamed. For The Naked Truth, when sitting in the café, she can control her public performance by hiding behind the pages of the book and pretending that her loneliness is an outcome of her own choice.

Thus, the performance as a single person that occurs under the observation of others can lead to the risk of embarrassment and shame—in other words, of *losing face* (Goffman 1967)—and attracting unwanted attention. Turning to Goffman again, the single woman who takes that risk can be seen as a kind of player in a ritual game, one who copes honorably or dishonorably, diplomatically or undiplomatically, with the judgmental contingencies of the situation (ibid., 31). Under these conditions, she cannot manage her presentation of the self and comply with temporal conventions. The Naked Truth's question also lends a new perspective to Jill Reynolds and Margaret Wetherell's (2003) argument that singleness is a discourse that orders particular subjectivities. Following this logic, it can be argued that such exposure can ascribe to her the stigmatized subjectivity of the lonely spinster without anyone to go out with.

The fear of appearing as a lonely single woman in public settings is a common concern. One can find plenty of instructional self-help books referring to this issue, or humorous videos aired on YouTube instructing one what to do when dining alone. In an episode of the popular television series *Friends*, aired on 1997, Rachel—one of the leading characters—tries to convince her friends to join her for dinner in a restaurant. Despite her efforts, she ends up dining alone. The episode depicts Rachel as particularly self-conscious about her aloneness in public. It is at that particular moment she bumps into a man she had made plans to go out with at a later date. The encounter is brief and awkward; but for Rachel, it is now clear that her future date thinks that she is "a total freak" by the mere act of dining by herself in a restaurant.

These representations demonstrate the importance of belonging to the right participation unit at the right time. This correlates with Goffman's claim that "a single is relatively vulnerable to contact, this being the grounds presumably why the ladies who inhabited traditional etiquette manuals did not appear in public unaccompanied; members of a with, after all, can count on some mutual protection" (Goffman 2010, 20–21). Following Goffman, we can reason that some of the interactional norms underlined in these behavioral and performative protocols render single women as particularly vulnerable to public scrutiny, as their status of being single is revealed as well as the fact that they have no one to count on for protection.

The privilege of civil inattention

The illustrations above provide vivid descriptions as to how single women are aware of the consequences of appearing alone in public. Goffman's analysis enables us to view the extent to which visibility and vulnerability in public are situational and relational to temporal codes of conduct. Returning to The Naked Truth's (2008) column, having coffee in a café is connected to day time activities, while going to a bar is related to evening and nighttime ones. As The Naked Truth elucidates, she can manage her self-presentation while having coffee, during the day, in a public café. The café at that time of the day provides her with protection and enables her interaction with the public to pass smoothly. Her participation unit as a *single* does not breach societal temporal norms, nor does it expose her status as a single woman. Her reflection attests that she feels less vulnerable during periods when she can control the public presentation of herself. As noted, this protection can be related to the time of the day, the week or the year. For instance, one of the prevailing assumptions that emerges here is that sitting during the day alone does not draw attention to her single status, while doing so at nighttime may trigger such attention.

It can be assumed that her solitary presence at nighttime prevents her from feeling part of the crowd. In contrast, when she goes to a bar with friends she can count on the mutual protection of the members of her participation unit and consequently blend in more easily without drawing unfavorable attention to herself. Thus, her decision to go out by herself during the day time and to avoid appearing by herself at certain times and in certain places is a strategic decision, which also deals with the significant question of agency and self-mastery. As she expresses it, in the café she can hide behind the pages of the book and pretend that her loneliness is an outcome of her own free choice.

These trivial, mundane, interactional rules can be dramatic for many single women. Dining out, going to the cinema, or attending a wedding, a family gathering, or a work-related activity require single women to strategically plan their appearance in public. Appearing at certain times as a single unit poses the risk of exposing her to what can be seen, in Goffmanian (1980) terms, as *uncivil attention,* as opposed to the *civil inat-tention* she is granted when one appears with a *with.*[2]

Goffman has observed that one of the commonplace strategies for retaining control of oneself in public is indifference to others. One often adopts modes of civil inatten-tion to strangers, granting them their own *personal space* while seeking to maintain one's own. Goffman has stressed that civil inattention is not inattention; on the con-trary, it recognizes the other's presence but it assumes respect for the other's personal space. Moreover, civil inattention is a manner in which people assess each other to gain knowledge and determine that the other does not pose any threat to them.

The injunction to appear as a *with* unit is also apparent in Byrne's study of single women in Ireland. As she notes:

> The theme of feeling isolated and being excluded by couples is reiterated by single women who explain the reluctance of others to include them in social gatherings. Because they are not partnered, they do not "fit in" or they lack shared interests with friends who are no longer single. Women's singleness is the problem. It is perceived by others as a problematic status and women are consequentially left out and leave themselves out of family and coupled gatherings. (Byrne 2000, 5)

As Byrne writes, single women feel isolated and excluded from family- and couple-oriented gatherings. Lalli, quoted above, formulates this as an exclusive party to which one has not been invited. The "closed party" metaphor, in this case, refers to her experiencing marginality in a couple-oriented society. However, she ends her column with these statements:

> I do have plans for this Valentine's Day: I will prepare soup, drink it with whole wheat bread and watch once again the film *Sliding Doors,* just to know that love is waiting for me. On second thoughts, perhaps we will meet here again next Valentine's Day. (Blue 2007)

The tone is similar to that of many texts written by single women, as they both conform and challenge dominant hegemonic traditional messages with regard to women's roles. However, Lalli's account also conveys a message which values the present, by making her own plans for Valentine's Day. These plans are not solely dependent upon belonging to the coupled unit.

In another column, Michal Shamir, a twenty-eight-year-old single woman discloses her discomfort of arriving unaccompanied to a work-related event:

> This year, my workplace posed a challenge which I cannot cope with: to show up with a date for a work-related event. This is the content of the email we received: "... For the very first time we are inviting your partner to celebrate with us the end of the summer happening!" ... I'm a twenty-eight-year-old single woman. I live in Tel Aviv, I'm independent and hard-working ... I love children and dogs ... yet I am still single without any relationship prospects ahead. This is the second time that my work poses a particular challenge: a work-related happening.
>
> As if it isn't enough to spend weekends alone—the most coupled time frame. As though Saturday mornings, Valentine's Day and the New Year's midnight kiss haven't done a good enough job in making my heart ache for the last five years; here comes the email from our dear CEO, reminding me that I have no one to bring to the event of the summer. All the other employees will be there with their spouses. Some are married and others are coupled. ... In two weeks it will happen, and probably I will deal with this again ... I still believe that life has a funny way of arranging itself. Anything can happen anywhere and at any time. Equipped with a smile ... once again, I will not give up. With or without a partner, I will be there. (Shamir 2012)

At the beginning of the column, Michal describes her working environment as pleasant and supportive. In such an environment, she is at ease and feels that she fits in. The corporate "fun day" represents an interruption to the everyday work routine, and in this sense sets up unexpected social boundaries between her and her partnered coworkers. Moreover, being a single woman, she has already begun to prepare her performance two weeks ahead of time. The invitation, to workers and their partners, conveys a clear message with regard to the expected interactional performance. The invitation to arrive as a "plus one" is considered as normative and natural in many social settings. This interactional etiquette is also commonplace in invitations to weddings and other public social occasions, copying the Noah's Ark pattern—in twos. As Michal notes, all the workers will be there with their partners. During everyday work time, she blends in and feels at ease in her working environment. However, the corporate fun day changes the status quo.

The plus one invitation poses a particular challenge for her as a single woman. On this occasion, the rules of social interaction change and accordingly her ability to self-master her performance in public alters. The invitation is an unexpected event, which leads her to reflect and re-evaluate her social status and consequently the participation unit she belongs to and is expected to belong to. It is worth noting that in these moments, single women sense that they are losing their individual agency and the ability to successfully manage their performance in public.

The different accounts describe encounters in which single women sense that their projected self is discredited in public. These fears and their attempts to keep their public composure give rise to strategic planning. One popular representation of this dynamic was depicted in Clare Kilner's film *The Wedding Date* (2005). In this romantic comedy, a single woman in her mid-thirties discovers that her younger sister is getting married. In order not to appear on her own at the wedding, she decides to pay for an escort to accompany her and save her from the public embarrassment of attending her younger sister's wedding as a single woman.

It could be argued that the Chinese e-commerce website mentioned at the beginning of this chapter has turned this comedic plot device into reality. It offers a date which enables Chinese single people to visit their parents for the New Year's holiday without losing face, and to comply accordingly with the social temporal expectancies demanded of them. The voices of the single women represented here understand perfectly well what Goffman (1959) has stressed regarding the importance of maintaining one's composure and succeeding in introducing favorable information about oneself.

Building on Goffman, this chapter has developed a dramaturgical perspective to time studies, showing how the shaping of one's identity and social relations is also temporally determined. Valentine's Day, New Year's Eve, the holiday seasons, wedding rituals, dinner or nighttime, the weekends: all are significant temporal markers, which play an important part in the lives of single women. In a couple- and familial-oriented world, they often determine single women's agentic capacities, and their hyper-visible yet invisible presence in public. As this chapter has shown, belonging to the appropriate and required participation unit becomes ever more crucial at particular times. Moreover, the temporal divisions between night and day, weekday and weekends, and work and leisure have significant bearing on the single woman's ability to successfully project a desired impression in public. The accounts, by and about single women, described here attest to their deep understanding of the social risks entailed in appearing in public unaccompanied, and how this leads to their heightened visibility and exposure as being single in a very couple-oriented world.

Notes

1   In Israel the weekend begins on a Friday.
2   For further discussion of uncivil attention, see also Gardner (1995) and Garland-Thomson (2009).

9

Discussion: another time

The very institutions that are directly responsible for much of the rigidity of our life—namely the Schedule and the Calendar—can also be seen as being among the foremost liberators of the modern individual. (Zerubavel 1985, 166)

In this quote, Zerubavel suggests that the calendar and the clock can also be among the "foremost liberators of the modern individual." As social actors we have more autonomy than we think and this includes re-articulating conventional temporal schemas and resisting heteronormative imperatives. Earlier in the book I wrote about Emma Morano and Jessie Gallan from Italy and Scotland who attributed their well-being and longevity to staying single for many years. Their stories have attracted global media attention and the *New York Times* piece about Morano was translated into many languages. In their interviews, both of them have expressed their contentedness and how they value their freedom and autonomy.

These stories invite an inquiry into the ways in which women's lives could exhibit and maintain an alternative temporality, one through which women can define their own past, present, and future and bestow it with their own rhythms and schedules. Morano's and Gallan's life stories present us with a way in which hegemonic social time can be destabilized and re-figured. Moreover, I suggest that their life-stories narrative subsumes a sense of controlling time, demonstrating their own life markers and temporal agency. Researching the web, one can find alternatives to the temporal regimes so rare within the conventional global romantic and familial cultural scripts. However, when one makes the effort to tap into different search categories, one can explore numerous internet sites, personal blogs, and local initiatives which seek to debunk common understandings and stereotypical attitudes towards single people. One example, an American website called Unmarried America, has earmarked a week in September as the "National Unmarried and Single Americans Week" in the US (Unmarried America 2015). Likewise, since around the early 2000s, self-help books have been available, such as *The Single Girl's Manifesta: Living in a Stupendously Superior Single State of Mind* (Stewart 2005), *Living Alone and Loving It* (Feldon 2003), and *Better Single than Sorry: A No Regrets Guide to Loving Yourself and Never Settling* (Schefft 2007).

When singlehood is represented as offering an alternative present and future, carving one's life on one's own terms without regrets is an act which has the potential to challenge the dominant scripts of the life directions women are expected to follow. Another site that seeks to challenge common-sense scripts is Sasha Cagen's website Quirkyalone. Cagen, the author of *Quirkyalone: A Manifesto for Uncompromising Romantics* (2004), coined the term Quirkyalone to present an alternative conceptualization to some of the prevalent stereotypes and images of single women. Quirkyalones, according to Cagen, are "People who enjoy being single (but are not opposed to being in a relationship) and generally prefer being single to dating for the sake of being in a couple" (Quirkyalone 2015).

These various articulations do not reiterate the pejorative images of long-term singlehood as a tragedy, nor do they subscribe to the still overwhelmingly heteronormative expectations directed at women. In opposition to the various statements and warnings analyzed over the course of this book ("You will die alone," "You will miss the train and stay on your own" etc.), the narratives that emerge from these cultural websites do not necessarily convey the regret of time wasted, or of missing out on the basic and essential experiences of life.

It does appear, however, that this counter-culture is significantly more developed in the UK and the US than in Israel. In Israel, I was unable to locate analogous initiatives on a comparable scale; indeed, it seems that there are no alternative Israeli "single-by-choice" websites and active bloggers who convey these messages. However, there are a few signs that indicate the possibility of a slow parallel transformation in Israel. For example, Rotem Lior, writing on the *Ynet* portal, introduces herself in the following way:

> Let me introduce myself: I am Rotem, a single woman not just by choice or a conscious decision but as a result of my very own will. Why? It's just in my nature. I prefer question marks to exclamation marks, expectation over certainty, and lust over statistics … Do people believe me? When they get to know me, they do. The only place where people do not believe me is the Internet. When I'm asked, in panic or in expectation, if I'm looking for a groom, I immediately declare that I'm a single woman. Here are some of the nicknames I have received in response: "poor thing" … "lonely liar," "lesbian," "feminist with too much hair," and "coward" … When they realize that you have passed the age of thirty, you can sense the rising suspicion. (Lior 2006)

Rotem Lior introduces lifelong singlehood here as a legitimate lifestyle option. However, she also emphasizes that this position is still rarely accepted and that it stirs profound disbelief and suspicion. Apparently, a woman who is over thirty and is still single necessarily implies individual defects: something is wrong. Nor does Lior embrace the time panic mode attached to single women above a certain age, through which one has to find a groom before it becomes too late. Thus her statement, "I'm a single woman," conveys a position in which she *lives in the present* and objects to the common perception of singlehood as a liminal, transitory position which is seen as betwixt and between. Thus, her simple declaration can be read as a way to reconstitute the present.

A similar stance is echoed in a column by Dvorit Shargal, a journalist and a documentarist, whose words I choose to analyze here in detail:

> It's four o'clock in the morning. Five o'clock in the morning. Six, Seven, Eight, Nine. Everything around me is quiet. There is no one snoring on the pillow next to me, no kids to take to kindergarten, no husband to drink my coffee with. Just me and, no one else but me. There is no organization which requires me to be part of it; everything is dependent on me, on my daily schedule, on the work which I have to finish, or on my training hour at the gym. And this peacefulness, this quietness of my life is my biggest happiness. This is an existential chosen static state which I wouldn't replace with any other noise. (Shargal 2006)

This portrayal of a typical morning also challenges the profound presupposition that singlehood is a temporary, non-chosen state. Dvorit stresses that her ever single status is an established and stable position that she has no intent of changing. Her position represents a personal and public identity of chosen singlehood which is based on volitional, autonomous decision-making. In this way, she opposes the widespread imagery that portrays single women as terrified by their imposed singlehood and expected loneliness. In other words, she dismantles the culturally constructed horror of sleeping alone, dining alone, or living alone, offering to replace them with images of contentedness and satisfaction. According to hegemonic hetero-temporalities, Shargal's time could be seen as *time on hold, meaningless and empty*. Yet, the writer stresses that she enjoys her solitary silence. The emptiness and stasis commonly ascribed to single time is configured, in her words, into a desirable schedule that could pose an alternative to "domestic bliss." Moreover, her status as an ever-single woman is not articulated in defensive and apologetic terms, and produces more channels for desires and longings.

In Chapter 5, I discussed how being single for "too long" can imply that one possesses an incompetent self. It might even label some as suffering from various deficiencies and pathologies. Drawing on Ahmed again, it can be deduced that norms of familial time are represented as a social good. Thus, the length and duration of family time is grasped as a *positive accumulation* of time to be praised and admired, while the accumulation of single time is configured as *wasted time*. If we also return to Thompson's (1967) well-known contention that the transition from task time to clock time turns time into a currency ruled by clock time; the longer a single woman is single, the more her exchange value decreases.

Dvorit's text echoes various accounts that have gained prominence across global media, and which convey a similar tone. These voices offer an alternative interpretation of being "off the market" altogether. Such a stance can be interpreted as refusing to conform to the ageist and sexist regulations of current dating practices. In chapter 5, I discussed how the temporal language of the "dating market place" is imbued with age-based schedules through which single women are objectified and evaluated. The "late single" or "ever single" option, when not engaged with a constant search for a partner, can pose an alternative to the oppressive discourses of heterosexuality and temporal market economy which so many daters accept as a given. Thus, by challenging or refusing to comply with the temporal rules of supply and demand, long-term single-hood can represent a free space within which one's value as a woman is not determined by one's exchange value and the judgmental scrutiny of men. In other words, lifelong singlehood can represent an option to refuse the control of the temporal regimes of beauty and youth. Claiming one's own *temporal autonomy*, rhythms, and schedules can pave a way towards defining women's time as not merely attuned to patriarchal and heteronormative dictates.

Thinking beyond the conjugal and familial imaginary presents such an alternative. Shargal's text continues to challenge the ways women are expected to use their time:

> I've always been alone. I am not a mother, neither am I married. I live my life within this big silence, which others are so afraid of. And why am I this way? The fact is I have no need to share my allotted time with other people. At least not in a sequential manner. That is, every now and then, here and there, for a couple of hours, that's fine. It is even desirable. But not more than that. (Shargal 2006)

Shargal claims her *temporal ownership*. She describes the benefits of being in charge of her schedules, which include being/not being with others. Silence is not perceived as empty or terrifying, but rather is desired and anticipated. Her domesticity is defined by different tempos and noises in which she prioritizes her *temporal autonomy*:

> More than not wanting any one to touch my personal stuff, I don't want anyone to inter-fere with my schedule. In fact, if I had to adapt myself to the schedules of others, I would not be able to do all the things that I can do today. If I had to raise children, I would not have time for myself. If I had to share my life with someone, I would wilt or wear away (ibid.)

Shargal presents us with alternative codes and alternative practices of time. She insists on establishing her own temporal routine, one disconnected from the dominant models of female time. Along these lines, these rhythms present us with the possibility of challenging the dominant understanding of a woman's home and her domestic gender roles. From this perspective, long-term singlehood can also be seen as an alter-native to prevailing conceptions of domesticity, and of women as household-family oriented consumers. In her landmark work *The Feminine Mystique*, Betty Friedan (1963) persuasively contended that the feminine mystique held that women could find fulfillment "through sexual passivity, acceptance of male domination, and nurturing motherhood" (73) According to Friedan, a significant component of this ideal was the perception of women as active consumers of home products, which are constantly purchased for their families.

I would therefore suggest that in this instance, singlehood can provide a feminist reading of domesticity and time by opening new ways of experiencing time, not regu-lated by the task of taking care of/shopping for others. Yet as I have argued before, female singlehood is not classless. It is important to note that this form of temporal ownership is contingent upon one's class location and material conditions. Acquiring temporal autonomy, and the ability to fulfill such a desire, is dependent on one's fiscal capacity to live on one's own. Being single can also be dependent upon one's material conditions. Thus, the ability of Shargal to have "no one interfere with her schedule" could be read as an outcome of the class privilege which grants her with such aptitude. It is beyond the scope of this study to develop this point further, but I hope that future studies of singlehood will address the multilayered intersections of class, time, and female singlehood.

This book raises the issue of the limited discursive resources available to long-term single living. I suggest that women like Shargal can offer alternative discursive resources and even claim their own symbolic capital by insisting on their own norms and set of priorities. Shargal's text takes me back to the current modalities of female time. Feminist scholars interested in the gendered dimensions of time have argued that women's time is perceived as relational. "Time is shared rather than personal and routinely experienced through the presence and expectations constituted in interpersonal relations (Odih 2007, xv). Because women's time is conventionally understood in relation to their roles as wives and mothers, it is no wonder that it is rare to counter such oppositional voices.

In a more recent column, Elinor Ferrara argues that singlehood is a choice, not a problem requiring a solution:

> One reason that I am a thirty-year-old single woman, God forbid, is free choice. Yes, yes, many women stay single for various reasons: this can be a desire to experience more relationships, meet more men instead of committing to one man and settling down. ... Who determines that we should all get married, have kids and buy a house with a crazy mortgage? Where does this obsessive desire emerge for manufacturing a uniform series of human beings who own a house and have a family? The sad part in all of this, is that many women (and people in general) do not know how to draw the line and tell the difference between what they really want and what their surroundings expect them to. (Ferrara 2013)

The notion of chosen singlehood, as we can see in Ferrara's account, can be deployed as an alternative discursive resource, from which assumptions that are taken for granted can be contested and refuted. I have discussed this issue more extensively elsewhere.[1] Here, I wish to outline again the limits of this discursive resource. Indeed, the right to choose stands at the heart of major contemporary feminist struggles. Choice can be practiced as justifying and encouraging resistance to hegemonic formations by seeking recognition for alternative ones. Identifying as a single by choice can pave the way for late singlehood or lifelong singlehood to be a legitimate and stable identity which offers counter-narratives to existing societal norms.

Yet the chosen singlehood discourse certifies binary modes of thinking, thereby establishing new hierarchies between those who can and cannot follow the dictates of the new regime of choice and self-monitoring. As these pronouncements suggest, choice should be seen as a discursive formula formed under socio-cultural conditions and contexts which limit and constrain these very choices. Moreover, and as I have clarified before, the new empowered images of liberated, freely choosing single women might essentialize women's lives and constitute new hierarchies between those who can and cannot follow the dictates of the new regime of choice and self-monitoring. If we wish to enrich our understandings of feminine singlehood, one should bear in mind that women's identities are connected to class, age, religion and sexual orientation, which enable and narrow one's options for holding on to the position of chosen singlehood. All that said, these new discourses should be taken seriously as they broaden our discursive and material horizons and subvert existing gender ideologies.

The various examples discussed here aim to debunk what are regarded as the defining aspects of femininity, and what is considered as worthy living. In this book I have tried to understand how these injunctions are conveyed through naturalized temporal norms and concepts. Single women like Shargal, Ferrara, Morano, Gallan, and many others bestow their time with meaning and a sense of direction, which do not necessarily cohere with familial narratives or lean upon the societal hierarchies distinguishing the coupled from the uncoupled. In that respect, they claim their own temporal agency by stressing their abilities to carve out their own schedules and timetables. Moreover, these timetables are not based entirely on hetero-patriarchal rhythms. Their time is not on hold, nor is it wasted or devalued. In this way, these accounts confuse dominant temporal perceptions and provide counter-perspectives, as well as offering new modalities of temporality, subjectivity, and social belonging.

Alternative life paths

Most of the texts analyzed in this book reflect the presumption that coupledom and family life promise enduring connectedness and meaningful sociability. However, recent studies conducted in the UK and the US amongst other countries show that in many cases, friends serve as a biographical anchor, and provide continuity and ontological security no less—and at times even more—than family life.[2] In a fascinating study about friendship ties, Shelly Budgeon (2006) argues that friends and non-familial relationships, as relationships of care, provide an important normative reference point in late modernity. The care and support of friendships and elective communities presents the possibility of organizing one's life trajectory outside, and not necessarily around, cohabiting couple relationships (Budgeon 2006; Roseneil 2004; Roseneil and Budgeon 2004). For many of the interviewees, friendships represent a significant source of continuity which provides on-going support (Budgeon 2006). Such an example can be found in Budgeon's interview with Carol, one of the respondents in her study, who had been considering a break up with her partner:

> It sounds awful to say it but if I were to put it on a balance sheet for what I get out of the relationship, I pay a heavy price for it. Whereas the friends that I have, I don't feel as though I pay a price there and so I would spend more time with people who give as much as I give them in a sense … It's not that I don't want to be in a relationship but this particular one, like my marriage before, I know it's run its course. It's not I want to be alone. It's just that neither of them were right for me and I think I've probably got a little bit cynical now and I think "well there isn't anybody that's right for you so you might as well just get on with your life and go out with friends and enjoy yourself and do what you want to do." (ibid.)

Carol is aware that, like her previous marriage, her relationship with her partner will eventually run its course. Like many women who have experienced divorce, the "happily ever after" ending convention does not fit with her life trajectory. Women like Carol sense that friendship ties can be more rewarding than conjugal ones, and prefer to allocate their time accordingly. A similar theme from Budgeon's study emerges from Jools' account in which she declares: "I think a friendship is for life, but I don't think a partner is … I'd marry my friends. They'd last longer" (ibid.)

Thus, one of the key findings in Roseneil and Budgeon's studies was the de-centering of couplehood in favor of friendship (Roseneil and Budgeon 2004). Friendship, according to Roseneil and Budgeon, emerges as a key relationship, offering the opportunity to disrupt heteronormative institutions and provide alternative life paths. These types of narratives, seldom heard in Israeli discourse, stand in contrast to what Roseneil rightly phrases as the "mythology of the singleton in desperate search for a marriage partner" (ibid., 413).

However the prevalence of warnings like "you will die alone," along with images of the lonely solitary spinster, downplay the role of friendships in the lives of many. Neither is friendship present in the horizon of linear reproductive time. Friends are mostly absent from the diagrams and charts of the life course which privilege conjugal and familial trajectories. Budgeon refers to Pahl's (2000) work on friendships, in which he claims that the only source of continuity is provided by friends, particularly when so many aspects of one's life may be transitory (jobs, marriages). As Pahl claims, "men and women may come to rely on their friends to provide support and confirmation of their enduring identities" (quoted in Budgeon 2006). Thus, friendships for the interviewees in Budgeon's study offer a greater degree of stability and continuity.

This analysis is important for rethinking and reconstructing conventional representations of the life course in which family life takes center stage. Long-term singlehood and friendship ties have no place or function in conventional life course charts. They are absent from the "happily ever after" scripts, based as they are on the promise of the happy couple and happy families. As singlehood scholars demonstrate, many single women maintain rich networks of friends through their life course. These relationships provide them with security and continuity (Simpson 2006; Trimberger 2005). In these studies, marriage emerges as a temporary phase, while friendship ties are the ones which can provide continuity and security. From this perspective, singlehood cannot be perceived as a liminal and transitory position, because marriage does not hold the ultimate path for "moving forward in life" and does not provide the connection between the present and future.

Another area which refutes the stigma of solitary aging are studies which underscore the importance of friendships in the lives of older women (Aday et al. 2006; O'Connor 1993, 1998). For example, a study about single women in a senior citizens' home stresses the lively network of social support that women develop and maintain. Indeed, it has been found that friendships increase one's morale, and may increase morale more than contact with family members (O'Connor, quoted in Aday et al.

2006). These kinds of messages are rarely heard and do not cohere with warnings such as "you will die alone," or with stereotypical images of the solitary and miserable old single woman, wandering alone, feeding cats etc.

A different way in which the supposedly tragic future of the single woman can be challenged is through the introduction of alternative models of single aging. In addition to the examples discussed at the beginning of this chapter, consider for example this short piece about *Dallas* star Linda Gray, published in an online preview to *HELLO!* magazine:

> Once left a loveless marriage—and to this day, she's perfectly content to live her life as a single woman. "A lot of women are content on their own and don't want a partner. They want their freedom," the actress, who has been largely single for the past 33 years, told *Hello!* "We all love to flirt. It doesn't mean you're going to bring the person home or have a sleepover" (*HELLO!* 2015)

This option of a seventy-five-year-old woman celebrity content to be "on her own" reveals a different social script to the hegemonic one. Gray's statement corresponds with recent research which reveals that many older women declare a preference for living alone rather than sharing their lives with men (and taking care of them) (Klinenberg 2012).

In August 2015, the publication of Kate Bolick's (2015) *Spinster: Making a Life of One's Own* garnered a lot of media attention. In her book, she writes about what she terms as her own *spinster wish*. Aware of the long legacy of spinsterhood as a derogatory term, she writes:

> I grant that a wholesale reclamation of the word spinster is a tall order. My aim is more modest: to offer it up as shorthand for holding on to that in you which is independent and self-sufficient, whether you're single or coupled.
>
> If you're single, whether never-married, divorced, or widowed, you can carry the word spinster like a talisman, a constant reminder that you're in very good company—indeed, part of a long and noble tradition of women past and present living on their own terms.
>
> If you find yourself unhappily coupled, you can use the word spinster to conjure a time when you weren't, and to recall that being alone is often far preferable to being in a bad relationship. (ibid., 296)

In Chapter 4, I explored the demonized "old maid" prototype, and outlined the various ways in which this stigmatic character is associated with what are perceived as marginal and a-synchronized temporalities. The commonplace image in mainstream culture of the desperate bridesmaid-to-be hoping to be next, or of the old maid sitting alone in her empty house surrounded by cats, marks the limited extent of the symbolic resources currently available to women. To a large extent, narratives which express a wish for solitude and the possible contentment of living on one's own are hardly ever presented. Reclaiming the pejorative term "spinster" represents in my view the next step that singlehood studies should take towards politicizing singlehood and challenging heteronormative time scales.

Politicizing singlehood

To a large extent, singlehood in Israel is a non-existent political category; it is neither identified nor included in the agendas of feminist, human rights, and social justice organizations. Furthermore, singlehood is very much underrepresented in critical studies curriculums at both the graduate and undergraduate level. In fact, to the best of my knowledge, challenging the discriminatory, patronizing attitudes towards late singlehood has been rarely on the agenda of any social or political organization. I therefore wish to ask why the discrimination and stigmatization against single persons has not translated to public and political initiatives. Based on this reality, I can point to some preliminary assumptions here, which explain why singlehood is yet to be politicized and become a target of feminist action.

First, the widely held perception of late or lifelong singlehood as a liminal, transitory phase and as a disruption and unnatural social category is very much alive and well in many societies. This line of thinking poses substantial obstacles for envisioning singlehood in political terms. Second, contemporary mainstream discourse mostly relates to singlehood through personal narratives, single women's life stories and advice columns. Thus, a complex web of discourses de-contextualizes singlehood from its wider social and cultural settings, leading to the widely held beliefs that attribute blame to single women themselves. Moreover, as I have shown, the widespread discourse not only puts the blame on single women, it emphasizes that their future will be nothing but a life of misery and loneliness. Paraphrasing Virginia Woolf's well-known dictum: she can perhaps live in "a room of her own but not in a house of her own."[3] Simply put, the message single women hear again and again is that they *cannot make it on their own.*

This is why accentuating the social and political dimensions of singlehood is an important step in the right direction. I suggest that the re-constitution of singlehood into a social category that one may wish to identify with—and form a political community with—can positively yield material and discursive changes. Here, I join DePaulo (2006), Reynolds (2008), and Moran (2004)[4] in their call for the politicization of singlehood and the need for a nuanced feminist engagement with the concept. This book is also a call for such needed intervention.

In this vein, some recent developments may inspire the hope of social change. At the time of writing, the 2016 American presidential election campaign was underway; media coverage of the campaign reflected what may lead to a significant change in the discourse about single women, and particularly the growing recognition of their voting power. Major newspapers such as the *The Economist*, the *Guardian*, the *New York Times*, and the *Washington Post* have dedicated extensive space to what is perceived as the potential and rising salience of single women in local and global politics

For example, a headline of *New York* magazine, from February 2016, declared that the single American woman had become "The most powerful voter this year, who in her rapidly increasing numbers has become an entirely new category of citizen" (Traister 2016b). Some of these discussions were triggered by the publication of a non-fiction book, already a best-seller, by Rebecca Traister (2016a), the writer of the *NY* article above. In the book, entitled *All the Single Ladies: Unmarried Women and the*

Rise of an Independent Nation, Traister notes that for the first time in history, unmarried women outnumber their married counterparts. She also argues that this state of affairs enables more women to pursue high-powered careers, and to live sexually diverse lives (ibid.).

It might be that these developments taking place in the US, alongside the growing numbers of single women worldwide, could lead the way to what I consider the next and required step in singlehood scholarship and advocacy. Moving forward could pave the way for encouraging both researchers and activists to become more involved in singlehood politics, perceiving singlehood in political terms, and attending to the unique needs of single persons. Thus far, relatively few scholars (DePaulo 2006; Hacker 2001) have vocalized the need to catalyze policy change for the single population.

The most prominent among them is Bella DePaulo, who is both a researcher and an activist. For the last two decades she has written for many years about how single persons are socially and economically discriminated against and do not enjoy the various financial benefits granted to couples and parents. She is one of the prominent advocates for this required change. In her scholarly works and numerous online columns and media interviews she promotes a new outlook on singlehood which views singlehood as a political consistency.

For example, even in a 2004 letter to the editor published in the *New York Times* opinion section, DePaulo makes several offers to the to the candidates running for presidency at that time:

1. Hit the books. Learn about the real place of singles in contemporary American society. Singles account for more than 40 percent of the electorate and work force. Households consisting of two parents and their children are slightly outnumbered by households comprised of a single person living alone. And most singles do not live alone. About nine million households are single-parent homes. Singles are also homeowners. Last year, they accounted for 46.7 percent of house sales. Singles are not predominantly youthful; only a third are aged 18 to 29. Singlehood is no longer a way station on the road to marriage. Women on average now spend more years of their adult lives single than married, and men are not far behind.

2. Learn the actual voting patterns. Despite the hype, it was not single women who had the lowest rate of voting in 2000, but single men. In their candidate preferences, the men stood out in their support of Ralph Nader (7 percent, compared to 4 percent for single women, and 2 percent for married men and women).

3. Master the issues of concern to singles. You will find, for example, that singles would like to make a decent living, have affordable health care and enjoy retirement. Their values are not antifamily—they are human values. The language of singles is the language of inclusiveness. Here is an example: "If you are willing to work hard and play by the rules, you are part of our family, and we're proud to be with you." It is from Bill Clinton's 1996 speech accepting the Democratic nomination for president. (DePaulo 2004)

I regard this letter as reflecting a new kind of politics that offers tangible possibilities for changing the public discourse and looks at singlehood as a visible community which political candidates have to take seriously. It also views single persons as political and citizen subjects with obligations and rights.

This also poses an alternative to the "family values" political discourse often conveyed by liberal, conservative, and even progressive parties. The rhetoric of "family values" or "ensuring our children's future" has come to stand for the public good, and of doing the right thing. Promoting issues and speaking on behalf of and for single persons is uncommon, if not inconceivable.

Another exception to this state of affairs was articulated by Raija Eeva, a Finnish politician and founder of the Finnish Association for Singles, quoted in one of DePaulo's online columns:

> An employer may purchase insurance for his or her employee. If the employee dies a claim will be paid out to a widow, widower or the employee's children. In the case of a single, the insurance company gets to pocket the claim. (DePaulo 2015)

In another article Eeva further argues:

> If I were to say that a social democrat or a Swedish-speaking person or an immigrant couldn't get the same tax rebate as someone else, there'd be a hue and cry. But apparently you can suffer injustice based on your legal or family status. (Yle 2014)

Such views express a confident call to end the discrimination against singles and the high price singles pay for their single status. That is, these voices do not accept and endorse hegemonic heteronormative practices of public acceptance. These suggested transformations are dependent upon changes in the public discourse of singlehood together with structural and institutional change.

It would be interesting to analyze the 2016 presidential campaign in the US, and to analyze its results taking into consideration the votes of single women and whether personal status affected their voting patterns. These new developments could lead to broad-based social and economic reforms, and the development of the material and discursive conditions that would encourage women to realize their agency.

In this context it is important to stress that one should take care not to relate to singlehood as one unitary category, and should distinguish between different types of representations of singles. This demands the consideration of, amongst other things, exogenous factors such as class, gender, religion, and race. Hopefully, this study can contribute to future research and thinking about ways to re-appropriate singlehood from its derogatory position and to remove its fixed connotations. As such, this book can be complemented by studies of the nuances and variances in the experiences and social contexts of women's singlehood.

Thus, my hope here is to re-conceptualise singlehood as a social and political category, which may in turn open more avenues for moving beyond the dichotomous and essentialist thinking of misery/happiness, togetherness/loneliness, and success/failure. Institutional policy-oriented reforms such as those proposed by Bella DePaulo (2006) and Daphna Hacker (2001) (building apartments designed for single households, or changing the tax structure for example) are important.[5] Yet they alone cannot create the new language and the discursive spaces necessary to rethink the conceptions of singlehood and familism so prominent in our widespread conventions of the worthy and good life.

Hopefully, this study can contribute to such an endeavor, by constructing and deconstructing some of the familiar and taken-for-granted meanings associated with singlehood. A different form of thinking on singlehood and time, one which explores and can envision alternative networks of support and solidarity while questioning the central place the family and couplehood occupy, might be a crucial first step in this direction.

Notes

1   I have written more about this topic in articles about the single woman's choice as a zero sum game. See Lahad (2014) and also in my analysis of singlehood and selectiveness, Lahad (2013). See also Dales (2005, 2014) for an interesting discussion of single women and agency in Japan.

2   See Budgeon 2006; Roseneil 2004; Roseneil and Budgeon 2004; Spencer and Pahl 2006; Trimberger 2005; Weston 1991.

3   For a fascinating analysis of South Korean single women's quest to acquire a room of their own, see Song (2010). For more discussions on the living arrangements of single women, see Dales (2013); Nakano (2011), Wilkinson (2014).

4   In her study *How Second Wave Feminism Forgot the Single Woman*, Moran (2004) claims that feminists have mostly (and successfully) lobbied for changes in education, employment, and reproductive rights. Thus, liberal feminism has mainly focused on reconciling work and family responsibilities, while the single woman remains neglected by the movement's agenda.

5   I refer readers once again to DePaulo's important and insightful columns, mostly appearing in her blog All Things Single (and More) http://belladepaulo.com/blog/, and also to her two other blogs: Living Single on the Psychology Today web portal www.psychologytoday.com/blog/living-single, and Single at Heart on the Psych Central web portal http://blogs.psychcentral.com/single-at-heart/. See also Arnold and Campbell (2013).

Bibliography

Adam, Barbara. 1990. *Time and Social Theory*. Cambridge: Polity Press.

———. 1995. *Timewatch: The Social Analysis of Time*. Cambridge: Polity Press.

Adams, Margaret. 1976. *Single Blessedness: Observations on the Single Status in Married Society*. New York: Basic Books.

Aday, Ronald H., Gayle C. Kehoe, and Lori A. Farney. 2006. "Impact of Senior Center Friendships on Aging Women Who Live Alone." *Journal of Women & Aging* 18 (1): 57–73.

Ahmed, Sara. 2010. *The Promise of Happiness*. Durham, NC: Duke University Press.

Althusser, Louis. 1971. "Ideology and Ideological State Apparatuses (Notes Towards an Investigation)." In *Lenin and Philosophy and Other Essays*. Translated by Ben Brewster, 127–186. London: Monthly Review Press.

Amir, Merav. 2007. "Bio-Temporality and Social Regulation: The Emergence of the Biological Clock." *Polygraph: An International Journal of Culture and Politics* 18: 47–72.

Arnold, Lisa and Christina Campbell. 2013. "The High Price of Being Single in America." *Atlantic*. January 14. Accessed April 5, 2013. www.theatlantic.com/sexes/archive/2013/01/the-high-price-of-being-single-in-america/267043/.

Aronoff, Myron J. 2000. "The 'Americanization' of Israeli Politics: Political and Cultural Change." *Israel Studies* 5 (1): 92–127.

Auyero, Javier. 2010. "Chuck and Pierre at the Welfare Office." *Sociological Forum* 25 (4): 851–860.

Avisror, Esti. 2011. "I'm 33 Years Old and My Uterus is Still Painfully Empty." *Ynet*. April 8. Accessed June 22, 2012. www.ynet.co.il/articles/0,7340,L-4053821,00.html.

Avraham, Eli and Anat First. 2003. " 'I Buy American': The American Image as Reflected in Israeli Advertising." *Journal of Communication* 53 (2): 282–299.

Bahadur, Nina. 2013. "Susan Patton, Princeton Mom, Is Back with More Advice." *Huffington Post*. June 22. Accessed July 18, 2014. www.huffingtonpost.com/nina-bahadur/susan-patton-princeton-mom-is-back_b_3133236.html.

Banosh, Noa. 2011a. "Hello Kitty: The Wacky Single Woman and Her Cat." *Ynet*. November 11. Accessed May 2, 2012. www.ynet.co.il/articles/0,7340,L-4146224,00.html.

———. 2011b. "A Single Woman at a Wedding: Just Don't Say to Me 'Bekarov Etzlech.' " *Ynet*. August 22. Accessed February 5, 2012. www.ynet.co.il/Iphone/Html/0,13406,L-Article-V7-4111936.html.

Bar, Dazy. 2009. "I Am Thirty Years Old, I Was Already Supposed to be Happily Married." *Ynet*. March 23. Accessed June 30, 2012. www.ynet.co.il/articles/0,7340,L-3689428,00.html.

Bartky, Sandra L. 1990. *Femininity and Domination: Studies in the Phenomenology of Oppression.* New York: Routledge.

Bat Chen. 2009. "Mother, I Am Sorry I Am Not Married Yet." *Ynet.* January 29. Accessed February 7, 2009. www.ynet.co.il/articles/1,7340,L-3662230,00.html.

Bauman, Zygmunt. 1992. *Mortality, Immortality and Other Life Strategies.* Stanford, CA: Stanford University Press.

———. 1996. "From Pilgrim to Tourist—or a Short History of Identity." In *Questions of Cultural Identity*, edited by Stuart Hall and Paul Du Gay, 18–36. London: Sage Publications.

———. 1998. *Globalization: The Human Consequences.* Cambridge: Polity Press.

———. 2000. *Liquid Modernity.* Cambridge: Polity Press.

———. 2003. *Liquid Love: On the Frailty of Human Bond.* Cambridge: Polity Press.

———. 2005. *Liquid Life.* Cambridge: Polity Press.

———. 2007. *Consuming Life.* Cambridge: Polity Press.

Bawin-Legros, Bernadette. 2004. "Intimacy and the New Sentimental Order." *Current Sociology* 52 (2): 241–250.

Beck, Ulrich and Elisabeth Beck-Gernsheim. 2002. *Individualization: Institutionalized Individualism and its Social and Political Consequences.* London: Sage Publications.

Becker, Gay. 1994. "Metaphors in Disrupted Lives: Infertility and Cultural Constructions of Continuity." *Medical Anthropology Quarterly* 8 (4): 383–410.

———.1999. *Disrupted Lives: How People Create Meaning in a Chaotic World.* Berkeley and Los Angeles, CA: University of California Press.

Becker, Howard S. 2008. *Outsiders.* New York: Simon & Schuster.

Beckett, Samuel. 1954. *Waiting for Godot.* New York: Grove Press.

Behrendt, Greg and Liz Tuccillo. 2009. *He's Just Not That into You: The No-Excuses Truth to Understanding Guys.* New York: Simon Spotlight Entertainment.

Bellah, Robert N., Steven M. Tipton, William M. Sullivan, Richard Madsen, and Ann Swidler. 1985. *Habits of the Heart: Individualism and Commitment in American Life.* Berkeley and Los Angeles, CA: University of California Press.

Ben-Porat, Guy. 2006. *Global Liberalism, Local Populism: Peace and Conflict in Israel/Palestine and Northern Ireland.* Syracuse, NY: Syracuse University Press.

Bennett, Keith C. and Norman L. Thompson. 1991. *Accelerated Aging and Male Homosexuality: Australian Evidence in a Continuing Debate.* New York: Harrington Park Press.

Berkovitch, Nitza. 1997. "Motherhood as a National Mission: The Construction of Womanhood in the Legal Discourse in Israel." *Women's Studies International Forum* 20 (5–6): 605–619.

Berlant, Lauren and Michael Warner. 1998. "Sex in Public." *Critical Inquiry* 24 (2): 547–566.

Bingham, John and Ashley Kirk. 2015. "Divorce Rate at Lowest Level in 40 Years after Cohabitation Revolution."*Telegraph.* November 23. Accessed April 25, 2016. www.telegraph.co.uk/news/uknews/12011714/Divorce-rate-at-lowest-level-in-40-years-after-cohabitation-revolution.html.

Bli K'chal Vesrak, Inbal. 2008. "I Shall Boycott Family Dinners Until I Find a Boyfriend." December 19. Accessed May 8, 2010. www.ynet.co.il/articles/0,7340,L-3634714,00.html.

Blue, Lalli. 2006. "Beauty Turns into Pain When You Have No One to Share It With." *Ynet.* November 12. Accessed March 15, 2008. www.ynet.co.il/articles/0,7340,L-3327107,00.html.

———. 2007. "For Those in Love, Every Day is a Day of Love—But What about Me?" *Ynet.* February 13. Accessed October 1, 2013. www.ynet.co.il/articles/0,7340,L-3363587,00.html.

Boellstorff, Tom. 2007. "When Marriage Falls: Queer Coincidences in Straight Time." *GLQ: A Journal of Lesbian and Gay Studies* 13 (2–3): 227–248.

———. 2008. *Coming of Age in Second Life: An Anthropologist Explores the Virtually Human.* Princeton: Princeton University Press.

Bolick, Kate. 2011. "All the Single Ladies." *Atlantic.* November. Accessed March 9, 2014. www.theatlantic.com/magazine/archive/2011/11/all-the-single-ladies/308654/.

———. 2015. *Spinster: Making a Life of One's Own.* New York: Penguin Random House.

Bourdieu, Pierre. 1993. "On the Family as a Realized Category." *Theory, Culture, and Society* 13 (3): 9–26.

———. 2000. *Pascalian Meditations.* Stanford, CA: Stanford University Press.

———. 2003. *Firing Back: Against the Tyranny of the Market 2.* New York: New Press.

Branigan, Tania. 2012. "New Year, New Fake Partners for China's Young Singletons with Parents to Please." *Guardian.* January 20. Accessed December 2, 2013. www.theguardian.com/world/2012/jan/20/new-year-fake-partners-china.

Brodsky-Kauffman, Esta. 2006a. "Bartender—Give Me a Double Shot of This." *nrg.* June 12. Accessed May 8, 2010. www.nrg.co.il/online/7/ART1/434/316.html.

———. 2006b. "Why Don't You Take Your Heart Out of the Refrigerator?" *nrg.* April 3. Accessed March 28, 2007. www.nrg.co.il/online/7/ART1/069/140.html.

———. 2007a. "Who Are You, Anyway?" *nrg.* October 29. Accessed December 18, 2014. www.nrg.co.il/online/55/ART1/652/383.html.

———. 2007b. "How to Get Rid of a Guy in Ten Seconds." *nrg.* April 23. Accessed November 22, 2009. www.nrg.co.il/online/7/ART1/572/433.html.

———. 2008a. "Boom? So That's It." *nrg.* March 14. Accessed March 16, 2014. www.nrg.co.il/online/55/ART1/709/323.html.

———. 2008b. "Stop Slipping to the Bottom Line." *nrg.* March 29. Accessed December 16, 2010. www.nrg.co.il/online/55/ART1/715/221.html.

———. 2009. "Trapped in the Net: On People Who Can't Rehabilitate from Dating Sites." *nrg.* August 28. Accessed July 15, 2010. www.nrg.co.il/online/55/ART1/932/374.html.

Buchmann, Marlis. 1989. *The Script of Life in Modern Society: Entry into Adulthood in a Changing World.* Chicago: University of Chicago Press.

Budgeon, Shelley. 2006. "Friendship and Formations of Sociality in Late Modernity: The Challenge of 'Post Traditional Intimacy.'" *Sociological Research Online* 11 (3). Accessed February 11, 2010. www.socresonline.org.uk/11/3/budgeon.html.

———. 2008. "Couple Culture and the Production of Singleness." *Sexualities* 11 (3): 301–325.

———. 2015. "The 'Problem' with Single Women: Choice, Accountability and Social Change." *Journal of Social and Personal Relationships.* Accessed October 10, 2015. doi: 10.1177/0265407515607647.

Bury, Michael. 1982. "Chronic Illness as Biographical Disruption." *Sociology of Health and Illness* 4: 167–182.

Bystrov, Evgenia. 2012. "Religion, Demography and Attitudes toward Civil Marriage in Israel 1969–2009." *Current Sociology* 60 (6): 751–770.

Byrne, Anne. 2000. "Singular Identities Managing Stigma, Resisting Voices." *Women's Studies Review* 7: 13–24.

———. 2003. "Developing a Sociological Model for Researching Women's Self and Social Identities." *European Journal of Women's Studies* 10 (4): 443–464.

———. 2009. "Perfidious and Pernicious Singlism." *Sex Roles* 60 (9–10): 760–763.

Byrne, Anne and Deborah Carr. 2005. "Caught in the Cultural Lag: The Stigma of Singlehood." *Psychology Inquiry* 16 (2–3): 84–91.

Cagen, Sasha. 2004. *Quirkyalone: A Manifesto for Uncompromising Romantics.* New York: Harper-Collins Publishers.

Central Bureau of Statistics. 2015. "Households—Selected Data—2013." Accessed February 6, 2016. www.cbs.gov.il/www/publications/meshek_bait/profil_mb.pdf .

Chambers, Deborah. 2001. *Representing the Family*. London: Sage Publications.

Chandler, Joan. 1991. *Women Without Husbands: An Exploration of the Margins of Marriage*. London: Macmillan.

Channel 2 News. 2010. "Do You Want to Rent a Date for the Holidays?" *Mako*. February 10. Accessed June 7, 2012. www.mako.co.il/news-world/international/Article-7ddca2eda08b621004.htm.

Charmaz, Kathy. 1997. *Good Days, Bad Days: The Self in Chronic Illness and Time*. New Brunswick, NJ: Rutgers University Press.

Chen, Moran. 2007. "I Am Only 21 and I Already Want to be Under the Bridal Canopy." *Ynet*. May 16. Accessed April 3, 2008. www.ynet.co.il/articles/0,7340,L-3399453,00.html.

China Daily. 2012. "Lonely Hearts Rent Dates to Please Parents." October 27. Accessed February 3, 2013. www.chinadaily.com.cn/china/2012-01/27/content_14495808.htm.

Cobb, Michael. 2012. *Single: Arguments for the Uncoupled*. New York: New York University Press.

Collins, Patricia H. 2000. *Black Feminist Thought: Knowledge, Consciousness, and the Politics of Empowerment*. New York: Routledge.

Cox, David. 2013. "Hong Kong's Women Are Suffering a Man Drought." *Nation*. December 7. Accessed January 20, 2014. www.nationmultimedia.com/opinion/Hong-Kongs-women-are-suffering-a-man-drought-30221496.html.

Craib, Ian. 1994. *The Importance of Disappointment*. London: Routledge.

Crapanzano, Vincent. 1985. *Waiting: The Whites of South Africa*. New York: Vintage.

Dagan, Chen. 2010. "The Or(a) [Light] at the End of the Tunnel: Ora Golan Releases Women from Chronic Singlehood." *nrg*. July 6. Accessed May 30, 2014. www.nrg.co.il/online/55/ART2/116/822.html.

Daily Mail. 2015. "Avoid Men and Eat Plenty of Porridge for a Long Life, Says Jessie, 109: Scotland's Oldest Woman Reveals Her Secrets." *Daily Mail*. January 15. Accessed February 10, 2015. www.dailymail.co.uk/news/article-2912299/Scotland-s-oldest-woman-Jessie-Gallan-reveals-longevity-secrets-including-eating-porridge-avoiding-men.html.

Dales, Laura. 2005. "Agency and the Parasite Single Issue." In *The Agency of Women in Asia*, edited by Kyn Parker, 133–157. Singapore: Marshall Cavendish Academic.

———. 2013. "Single Women and their Households in Contemporary Japan." In *The Global Political Economy of the Household in Asia*, edited by Juanita Elias and Samanthi J. Gunawardana, 110–126. Basingstoke: Palgrave Macmillan.

———. 2014. "Ohitorisama, Singlehood and Agency in Japan." *Asian Studies Review* 38 (2): 224–242.

Daly, Kerry J. 1996. *Families and Time: Keeping Pace in a Hurried Culture*. Thousand Oaks, CA: Sage Publications.

Dates4Hire. 2014. "About Us: Hire a Date—Rent a Date for Any Occasion—Get Paid to Date." Accessed March 15, 2015. www.dates4hire.com/about-us.

Davidovitz, Dana. 2010. "Dates Fasting, Cleansing the Body in Preparation for Love." *Ynet*. April 16. Accessed March 19, 2014. www.ynet.co.il/articles/1,7340,L-3876786,00.html.

Davies, Madlen. 2015. "The Secret to Living past 100? Stay Single and Eat Three Raw Eggs a Day, Says Europe's Oldest Woman." *Daily Mail*. February 17. Accessed March 1, 2015. www.dailymail.co.uk/health/article-2957124/The-secret-living-past-100-Stay-single-eat-three-raw-eggs-day-says-Europe-s-oldest-woman.html#ixzz3SJUpPRBc.

Davies, Michele L. 1997. "Shattered Assumptions: Time and the Experience of Long-Term HIV Positivity." *Social Science & Medicine* 44 (5): 561–571.

Davies, Karen. 1994. "The Tensions between Process Time and Clock Time in Care-Work: The Example of Day Nurseries." *Time & Society* 3 (3): 277–303.

DePaulo, Bella M. 2004. "Sex and the Single Voter." *New York Times.* June 18. Accessed November 13, 2013. www.nytimes.com/2004/06/18/opinion/18DEPA.html.

———. 2006. *Singled Out: How Singles are Stereotyped, Stigmatized, and Ignored, and Still Live Happily Ever After.* New York: St. Martin's Press.

———. 2008. "Holiday Spirit, 21st Century Style: Kay Trimberger and I Share Our Vision." *Living Single* (blog), November 23. Accessed June 25, 2012. www.psychologytoday.com/blog/living-single/201009/did-second-wave-feminism-neglect-the-single-woman.

———. 2010. "Did Second-Wave Feminism Neglect the Single Woman?" *Living Single* (blog), September 5. Accessed July 9, 2013. www.psychologytoday.com/blog/living-single/201009/did-second-wave-feminism-neglect-the-single-woman.

———. 2015. "Singlism in Finland is Causing a Stir." *Single at Heart* (blog). Accessed January 26. Accessed January 19, 2016. http://blogs.psychcentral.com/single-at-heart/2014/06/singlism-in-finland-is-causing-a-stir/#comments.

DePaulo, Bella M. and Wendy L. Morris. 2005. "Singles in Society and in Science." *Psychological Inquiry* 16 (2–3): 57–83.

Domkeh, Ronit. 2014. "I'm a 35 Year Old Single Woman. So What?" *The Marker.* April 30. Accessed September 6, 2015. www.themarker.com/magazine/1.2289538.

Doron, Yael. 2010. "Fed Up of Being 35 Years Old and Feeling Old and Desperate." *Ynet.* September 8. Accessed November 1, 2012. www.ynet.co.il/articles/0,7340,L-3951556,00.html.

Doron, Yael and Gili Bar. 2007. "7 Soul Searchings for Valentines Day." *Ynet.* July 30. Accessed February 5, 2008. www.ynet.co.il/articles/0,7340,L-3430982,00.html.

———. 2008. "I Am Afraid to End Up Like My Grandmother and My Mother, Who Married When They Were Young." *Ynet.* May 15. Accessed July 5, 2014. www.ynet.co.il/articles/0,7340,L-3539746,00.html.

———. 2009. "My Wife Feels She is Missing Something on the Outside." *Ynet.* March 3. Accessed December 8, 2010. www.ynet.co.il/articles/0,7340,L-3680055,00.html.

Durkheim, Émile. 2008. *The Elementary Forms of the Religious Life.* Mineola, NY: Dover Publications.

Donath, Orna. 2011. *Making a Choice: Being Child-Free in Israel.* Tel Aviv: Miskal-Yedioth Ahronoth Books.

———. 2015 "Regretting Motherhood: A Sociopolitical Analysis." *Signs: Journal of Women in Culture and Society* 40 (2): 343–367.

Donovan, J. (Ed.) 1995. *Animals and Women: Feminist Theoretical Explorations.* Durham: Duke University Press.

Ebaugh, Helen R. F. 1988. *Becoming an Ex: The Process of Role Exit.* Chicago: University of Chicago Press.

Eck, Beth A. 2013. "Identity Twists and Turns: How Never-Married Men Make Sense of an Unanticipated Identity." *Journal of Contemporary Ethnography* 42 (1): 31–63.

———. 2014. "Compromising Positions: Unmarried Men, Heterosexuality, and Two-Phase Masculinity." *Men and Masculinities* 17 (2):147–172.

Elias, Norbert. 1992. *Time: An Essay.* Oxford: Blackwell Publishing.

Engelberg, Ari. 2011. "Seeking a 'Pure Relationship'? Israeli Religious-Zionist Singles Looking for Love and Marriage." *Religion* 41 (3): 431–448.

———. (2013). "Religious Zionist Singles and Late-Modern Youth Culture." *Israel Studies Review,* 28 (2): 1–17.

Evertsson, Lars and Charlott Nyman. 2013. "On the Other Side of Couplehood: Single Women in Sweden Exploring Life without a Partner." *Families, Relationships and Societies* 2 (1): 61–78.

Farkash, Shirli. 2007. "A Thousand Imaginary Friends No Longer Helped, I Felt Alone." *Ynet.* March 2. Accessed March 22, 2008. www.ynet.co.il/articles/0,7340,L-3371239,00.html.

Farkol, Sarit. 2007. "After a Week, They Have Decided to Live Together." *Ynet.* September 9. Accessed April 21, 2008. www.ynet.co.il/articles/0,7340,L-3447218,00.html.

Feldon, Barbara. 2003. *Living Alone and Loving it.* New York: Fireside.

Ferrara, Elinor. 2013. "Singlehood is a Choice! Not a Problem That Needs Solving." *Ynet.* April 23. Accessed May 2, 2013. www.ynet.co.il/articles/0,7340,L-4371372,00.html.

Finch, Janet. 2007. "Displaying Families." *Sociology* 41 (1): 65–81.

Fiske, John. 1996. *Media Matters: Race and Gender in US Politics.* Minneapolis: University of Minnesota Press.

Fogiel-Bijaoui, Sylvie. 1999. "Families in Israel: Between Familialism and Post Modernity." In Dafna Izraeli et al. (eds.), *Gender, Sex, Politics*, pp. 107–166. Tel Aviv: Am Oved.

———. 2002." Familism, Postmodernity and the State: The Case of Israel." *The Journal of Israeli History* 21 (1–2): 38–62.

Fogiel-Bijaoui, Sylvie and Reina Rutlinger-Reiner. 2013. "Rethinking the Family in Israel." *Israel Studies Review* 28 (2): vii–xii.

Frankfort-Nachmias, Chava and Erella Shadmi. 2004. *Sappho in the Holy Land: Lesbian Existence and Dilemmas in Contemporary Israel.* Albany: State University of New York Press.

Friedan, B. 1963. *The Feminine Mystique.* New York: W. W. Norton & Company.

Friedman, Hadas. 2007. "Summer Love and One Last Kiss." *Ynet.* August 30. Accessed March 5, 2013. www.ynet.co.il/Ext/Comp/ArticleLayout/CdaArticlePrintPreview/1,2506, L-3443659,00.html.

———. 2009. "A 35-Year-Old Single Woman: The Stigma." *Ynet.* March 1. Accessed April 10, 2015. www.ynet.co.il/articles/0,7340,L-3678745,00.html.

Foucault, Michel. 1972. *The Archaeology of Knowledge and the Discourse on Language.* New York: Pantheon Books.

———. 1991. *Discipline and Punish: The Birth of the Prison.* Harmondsworth: Penguin.

Franklin, Sarah. 1997. *Embodied Progress: A Cultural Account of Assisted Conception.* London: Routledge.

Gaetano, Arianne. 2009. "Single Women in Urban China and the 'Unmarried Crisis': Gender Resilience and Gender Transformation." Working Paper no. 31, Center for East and South-East Asian Studies, Lund University, Sweden. http://lup.lub.lu.se/luur/download?func=downloadFile&recordOId=3127765&fileOId=3127772.

Gal, Orit. 2007. "Do You Want Love Like in the Movies or Do You Want to Get Married?" *Ynet.* September 2. Accessed January 18, 2015. www.ynet.co.il/articles/0,7340,L-3444527,00.html.

———. 2010. "Interestingly There is No Stigma of an Aging Single Man." *Ynet.* April 12. Accessed July 20, 2011. www.ynet.co.il/articles/0,7340,L-3427587,00.html.

Gardiner, Judith K. 2002. "Theorizing Age with Gender: Bly's Boys, Feminism and Maturity Masculinity. " In *Masculinity Studies and Feminist Theory: New Directions*, 90–118. New York: University of Columbia Press.

Gardner, Carol. 1995. *Passing By: Gender and Public Harassment.* Berkeley and Los Angeles, CA: University of California Press.

Garland-Thomson, Rosemarie. 2002. "Integrating Disability, Transforming Feminist Theory." *NWSA Journal* 14 (3): 1–32.

———. 2009. *Staring: How We Look.* New York: Oxford University Press.

Garner, J. Dianne. 1999. "Feminism and Feminist Gerontology." In *Fundamentals of Feminist Gerontology*, edited by J. Dianne Garner, 3–12. New York: Haworth Press.

Gasparini, Giovanni. 1995. "On Waiting." *Time & Society* 4 (1): 29–45.

Gearan, Anne. 2007. "Rice: Single Women Can Understand War." *USA Today*. January 14. Accessed March 15, 2014. http://usatoday30.usatoday.com/news/washington/2007-01-14-rice-war_x.htm.

Giddens, Anthony. 1990. *The Consequences of Modernity*. Stanford, CA: Stanford University Press.

———. 1991. *Modernity and Self-Identity: Self and Society in the Late Modern Age*. Stanford, CA: Stanford University Press.

———. 1992. *The Transformation of Intimacy: Sexuality, Love and Eroticism in Modern Societies*. Cambridge: Polity Press.

Giga. 2007. "I Am 36 and I Too Want to Be Married." *Ynet*. February 16. Accessed February 3, 2008. www.ynet.co.il/articles/0,7340,L-3365916,00.html.

Glaser, Barney G. and Anselm L. Strauss. 1971. *Status Passage*. Chicago: Aldine-Atherton.

Goffman, Erving. 1959. *The Presentation of Self in Everyday Life*. New York: Anchor Books.

———. 1961. *Asylum: Essays on the Social Situation of Asylum Patients and Other Inmates*. Garden City, NJ: Anchor Books.

———. 1967. *Interaction Ritual*. New York: Pantheon Books.

———. 1969. *Strategic Interaction*. Philadelphia: University of Pennsylvania Press.

———. 1972. "The Moral Career of the Mental Patient." In *Symbolic Interaction*, edited by Jerome G. Manis and Bernard N. Meltzer, 234–244. Boston, MA: Allyn & Bacon.

———. 1980. *Behavior in Public Places: Notes on the Social Organization of Gatherings*. Westport, CT: Greenwood Press.

———. 1983. "The Interaction Order: American Sociological Association, 1982 Presidential Address." *American Sociological Review* 48 (1): 1–17.

———. 2010. *Relations in Public: Microstudies of the Public Order*. New Brunswick, NJ: Transaction Publishers.

Goldenberg, Roe. 2015. "Where Did We Surf: What Are the Most Popular Websites in Israel?" *Globes*, January 13. Accessed February 20, 2015. www.globes.co.il/news/article.aspx?did=1001000487.

Goldstein-Gidoni, Ofra. 2012. *Housewives of Japan: An Ethnography of Real Lives and Consumerized Domesticity*. Basingstoke: Palgrave Macmillan.

Gooldin, Sigal. 2008. "Technologies of Happiness: Fertility Management in a Pro-Natal Context." In *Citizenship Gaps: Migration, Fertility and Identity in Israel*, edited by Yossi Yonah and Adriana Kemp. Jerusalem: Van-Leer Jerualem Institute/Hakibutz Hameuchad Publishing House: 167–206.

Gordon, Tuula. 1994. *Single Women: On the Margins?* New York: New York University Press.

Gottlieb, Lori. 2008. "Marry Him! The Case for Settling for Mr. Good Enough." *Atlantic*. March. Accessed October 25, 2010. www.theatlantic.com/magazine/archive/2008/03/marry-him/306651/.

Greenhouse, Carol J. 1996. *A Moment's Notice: Time Politics Across Cultures*. Ithaca, NY: Cornell University Press.

Greenwald, Rachel. 2004. *Finding a Husband After Thirty-Five: Using what I Learned at Harvard Business School*. New York: Ballantine Books.

Gullette, M. M. (1998). "Midlife Discourses in the Twentieth-Century United States: An Essay on the Sexuality, Ideology, and Politics of 'Middle-Ageism.'" *Welcome to Middle Age*, 3–44.

———. 2004. *Aged by Culture*. Chicago: University of Chicago Press.

Hacker, Daphna. 2001. "Single and Married Women in the Law of Israel—a Feminist Perspective." *Feminist Legal Studies* 9 (1): 29–56.

———. 2004. "Beyond Old Maid and Sex and the City: Singlehood as an Important Option for Women and Israeli's Law Attitude Towards This Option." *Tel Aviv University Law Review* 28 (3): 903–950.

Hage, Ghassan (Ed.) 2009. *Waiting.* Melbourne, VIC: Melbourne University Press.

Halberstam, Judith. 2005. *In a Queer Time and Place: Transgender Bodies, Subcultural Lives.* New York and London: New York University Press.

———. 2011. *The Queer Art of Failure.* Durham, NC: Duke University Press.

Hall, Stuart. 2010. "Notes on Deconstructing 'the Popular.'" In *Cultural Theory: An Anthology,* edited by Imre Szeman and Timothy Kaposy, 72–80. Malden, MA: John Wiley & Sons.

Handelman, Don. 2004. *Nationalism and the Israeli State: Bureaucratic Logic in Public Events.* New York: Berg Publishers.

Hanick Zikukit Tafus. 2011. *"Bekarov Etzlech." A Little Bit of That and a Little Bit of That and That … (Tapuz),* June 5. Accessed May 5, 2012. www.tapuz.co.il/blog/net/ViewEntry.aspx?entryId=2011911&skip=1.

Hardy, Chrissa. 2015. "Meet Emma Morano, the 115-Year-Old Woman Who Says Staying Single is the Secret to a Long & Happy Life." *Bustle.* Accessed April 1, 2016. www.bustle.com/articles/64474-meet-emma-morano-the-115-year-old-woman-who-says-staying-single-is-the-secret-to-a-long.

Harel, Alon. 1999. "The Rise and Fall of the Israeli Gay Legal Revolution." *Columbia Human Rights Law Review* 31: 443–471.

Hart, Nicky. 1976. *When Marriage Ends: A Study in Status Passage.* London: Tavistock Publications.

Hashachar, Tal. 2011. "Not Pretty and Not Ugly: You Are Ok, They Like Average." *Ynet.* August 4. Accessed May 20, 2012. www.ynet.co.il/articles/0,7340,L-4104140,00.html.

Hashiloni-Dolev, Yael. 2007. *A Life (Un)Worthy of Living: Reproductive Genetics in Israel and Germany.* Dordrecht: Springer.

Hashiloni-Dolev, Yael and Shiri Shkedi. 2007. "On New Reproductive Technologies and Family Ethics: Pre-Implantation Genetic Diagnosis for Sibling Donor in Israel and Germany." *Social Science & Medicine* 65 (10): 2081–2092.

Hazan, Haim. 1980. *The Limbo People: A Study of the Constitution of the Time Universe Among the Aged.* London: Routledge & Kegan Paul.

———. 1994. *Old Age: Constructions and Deconstructions.* Cambridge: Cambridge University Press.

———. 2002. "Dis-Membered Bodies—Re-Membered Selves: The Discourse of the Institutionalized Old." In *Cultural Gerontology,* edited by Lars Andersson, 207–220. Westport, CT: Auburn House.

———. 2006. "Age." In *Equality,* edited by Uri Ram and Nitza Berkovitch, 52–56. Be'er Sheva: Ben Gurion University Press.

Heaphy, Brian. 2011. "Critical Relational Displays." In *Displaying Families: A New Concept for the Sociology of Family Life,* edited by Julie Seymour and Andrea Doucet, 19–37. Basingstoke: Palgrave Macmillan.

Heart, Goldy. 2008. "Passover, Soon You Will Hear the *Bekarov Ezlech.*" *Ynet.* April 14. Accessed November 1, 2013. www.ynet.co.il/articles/0,7340,L-3531967,00.html.

Heather, Ben and Paul Easton. 2014. "Man Drought Leaves Many Lacking Romance." *Stuff.* January 11. Accessed December 19, 2014. www.stuff.co.nz/life-style/love-sex/9598458/Man-drought-leaves-many-lacking-romance.

Heimtun, Bente. 2010. "The Holiday Meal: Eating Out Alone and Mobile Emotional Geographies." *Leisure Studies* 29 (2): 175–192.

———. 2012. "Proposing Paradigm Peace: Mixed Methods in Feminist Tourism Research." *Tourist Studies* 12 (3): 287–304.

HELLO! 2015. "Linda Gray Speaks Exclusively to *HELLO!* About Her Ex-Husband and How Dallas Changed Her Life." September 21. Accessed October 22, 2015. www.hellomagazine.com/celebrities/2015092127289/linda-gray-talks-exclusively-to-hello-about-end-of-marriage/.

Herbst, Anat. 2013. "The Legitimacy of Single Mothers in Israel Examined through Five Circles of Discourse." *Israel Studies Review* 28 (2): 228–246.

Hidas, Yuval. 2015. "Egg Freezing: Insurance Certificate That Promise Nothing, Only an Option, Perhaps." Unpublished Master's thesis, Tel Aviv University.

Hine, Christine. 2000. *Virtual Ethnography*. London: Sage Publications.

Hochschild, Arielle. 2003. *The Commercialization of Intimate Life: Notes from Home and Work*. Berkeley and Los Angeles: University of California Press.

Holstein, James A. and Jaber F. Gubrium. 2000. *Constructing the Life Course*. New York: General Hall.

Horowitz, Dan and Moshe Lissak. 1989. *Trouble in Utopia: The Overburdened Polity of Israel*. Albany, New York: State University of New York Press.

Howell, Signe. 2001. "Self-Conscious Kinship: Some Contested Values in Norwegian Transnational Adoption." In *Relative Values: Reconfiguring Kinship Studies*, edited by Sara Franklin and Susan McKinnon, 203–223. Durham, NC: Duke University Press.

Hughes, Everett C. 1971. *The Sociological Eye: Selected Papers on Work, Self & the Study of Society*. Chicago: Aldine-Atherton.

Illouz, Eva. 1997. *Consuming the Romantic Utopia: Love and the Cultural Contradictions of Capitalism*. Berkeley and Los Angeles: University of California Press.

———. 2003. *Oprah Winfrey and the Glamour of Misery*. New York: Columbia University Press.

———. 2007. *Cold Intimacies: The Making of Emotional Capitalism*. Cambridge: Polity Press.

Ingraham, Chrys. 1999. *White Weddings: Romancing Heterosexuality in Popular Culture*. New York: Routledge.

Inhorn, Rosy and Sari Zimmerman. 2007. "Many of My Friends are Getting Divorced, Why Should I Get Married?" *Ynet*. December 6. Accessed March 15, 2014. www.ynet.co.il/articles/0,7340,L-3411367,00.html.

Israel Ministry of Health. 2011. "Egg Freezing for the Purpose of Preserving the Woman's Fertility." Circular no. 1/2011. January 9. Accessed May 17, 2013. www.health.gov.il/hozer/mr01_2011.pdf.

Jahoda, Marie and Hans Zeisel. 1974. *Marienthal: The Sociography of an Unemployed Community*. NJ. Transaction Publishers.

Jamieson, Lynn and Roona Simpson. 2013. *Living Alone: Globalization, Identity and Belonging*. London: Palgrave Macmillan.

Jeffrey, Craig. 2010. *Timepass: Youth, Class, and the Politics of Waiting in India*. Stanford, CA: Stanford University Press.

Jones, Julie and Steve Pugh. 2005. "Ageing Gay Men: Lessons from the Sociology of Embodiment." *Men and Masculinities* 7 (3): 248–260.

Kahn, Susan M. 2000. *Reproducing Jews: A Cultural Account of Assisted Conception in Israel*. Durham and London: Duke University Press.

Katz-Gerro, Tally, Sharon Raz, and Meir Yaish. 2009. "How Do Class, Status, Ethnicity, and Religiosity Shape Cultural Omnivorousness in Israel?" *Journal of Cultural Economics* 33 (1): 1–17.

Keith, Hamish and Dinah Bradley. 1991. *Becoming Single: How to Survive When a Relationship Ends*. East Roseville, NSW: Simon & Schuster Australia.

Kesler, Dana. 2012. "Stop on the Way to the Wedding: Single People in Crisis." *nrg*. July 27. Accessed August 2, 2014. www.nrg.co.il/online/1/ART2/390/117.html.

Kimchi, Adi. 2013. "Sylvester Alarm: Finding a Date for Tonight." *Ynet*. December 31. Accessed March 2, 2014. www.ynet.co.il/articles/0,7340,L-4471594,00.html.

———. 2014. "Passover and You are Alone: How to Cope with Your Family." *Ynet*. April 10. Accessed December 3, 2014. www.ynet.co.il/articles/0,7340,L-4508976,00.html.

Klinenberg, Eric. 2012. *Going Solo: The Extraordinary Rise and Surprising Appeal of Living Alone*. New York: Penguin Press.

Kohli, Martin. 2007. "The Institutionalization of the Life Course: Looking Back to Look Ahead." *Research in Human Development* 4 (3–4): 253–271.

Kohli, Martin and John W. Meyer. 1986. "Social Structure and Social Construction of Life Stages." *Human Development* 29 (3): 145–149.

Kremba, Lotti. 2009. "Thirty-Nine Years Old—You have Missed the Men's Train." *Ynet*. January 21. Accessed August 4, 2012. www.ynet.co.il/articles/0,7340,L-3659552,00.html.

Kuchler, Hannah and Emma Jacobs. 2014. "Egg-Freezing is Latest Talking Point in Valley Diversity Debate." *Financial Times*. October 17. Accessed December 5, 2014. www.ft.com/intl/cms/s/0/fc32c6c6-5599-11e4-89e8-00144feab7de.html#axzz3VyCM1yPk.

Lahad, Kinneret. 2007. "To Clarify, to Help and to Solve: On Self Help Books for Singles." *Theory and Criticism* 30: 238–249.

———. 2013. "'Am I Asking For Too Much?' The Selective Single Woman as a New Social Problem." *Women's Studies International Forum* 40: 23–32.

———. 2014. "The Single Woman's Choice as a Zero-Sum Game." *Cultural Studies* 28 (2): 240–266.

Lahad K. and A. Shoshana (2015). "Singlehood in *Treatment*: Interrogating the Discursive Alliance Between Postfeminism and Therapeutic Culture. *European Journal of Women's Studies* 22(3): 334–349.

Lakoff, George and Mark Johnson. 1980. "Conceptual Metaphor in Everyday Language." *The Journal of Philosophy* 77 (8): 453–486.

Landau, Ruth. 1996. "Assisted Reproduction in Israel and Sweden: Parenthood at Any Price?" *International Journal of Sociology and Social Policy* 16 (3): 29–46.

Laz, Cheryl. 1998. "Act Your Age." *Sociological Forum* 13 (1): 85–113.

Lazar, Michelle M. 2007. "Feminist Critical Discourse Analysis: Articulating a Feminist Discourse Praxis." *Critical Discourse Studies* 4 (2): 141–164.

Leach, Edmund. 1971. *Rethinking Anthropology*. London: Athlone Press.

Lefebvre, Henri. 2004. *Rhythmanalysis: Space, Time and Everyday Life*. London: Continuum.

Levi, Hezi. [Chief of Medical Administration.] 2011. Memorandum to general hospital directors and HMOs' medical departments directors. "Egg Freezing in Order to Preserve the Woman's Fertility." January 9. Accessed May 10, 2015. www.health.gov.il/hozer/mr01_2011.pdf.

Levin, May. 2006. "Everyone Thinks They Are Miki Buganim [An Israeli Stylist]." *Ynet*. May 18. Accessed May 26, 2013. www.ynet.co.il/articles/0,7340,L-3252560,00.html.

Levinson, Daniel J. 1978. *The Seasons of a Man's Life*. New York: Knopf.

Lior, Rotem. 2006. "A Thirty-Plus-Year-Old Woman, Happy and Single, Thank You Very Much." *Ynet*. June 14. Accessed April 10, 2010. www.ynet.co.il/articles/0,7340,L-3261626,00.html.

———. 2007. "'The Beautiful and The Brave': This Is How You Will Find Love." *Ynet*. April 4. Accessed April 5, 2008. www.ynet.co.il/articles/0,7340,L-3384311,00.html.

Look4Love. 2015. "Speed Date—Efficient Dating." Accessed December 27, 2015. www.look4love.co.il/AboutSpeedDate.asp.

Louise. 2007. "That's It! I'm Not Going Out on Any More Dates!" *Ynet*. June 15. Accessed February 20, 2009. www.ynet.co.il/articles/0,7340,L-3413048,00.html.

Lustenberger, Sibylle. 2013. Conceiving Judaism: The Challenges of Same-Sex Parenthood. *Israel Studies Review* 28 (2): 140–56.

Lyman, Stanford M. and Marvin B. Scott. 1989. *A Sociology of the Absurd*. New York: General Hall.

MacVarish, Jan. 2006. "What is the Problem of Singleness?" *Sociological Research Online* 11 (3). Accessed May 25, 2013. www.socresonline.org.uk/11/3/macvarish.html.

Maeda, Eriko and Michael L. Hecht. 2012. "Identity Search: Interpersonal Relationships and Relational Identities of Always-Single Japanese Women over Time." *Western Journal of Communication* 76 (1): 44–64.

Malachy, Shirli. 2010. "I Feel I'm Missing Out on My Life." *Ynet*. July 15. Accessed July 18, 2014. www.ynet.co.il/articles/0,7340,L-3920283,00.html.

Mann, Leon. 1969. "Queue Culture: The Waiting Line as a Social System." *American Journal of Sociology* 75 (3): 340–354.

Mazzarella, Sharon R. 2007. "Cyberdating Success Stories and the Mythic Narrative of Living 'Happily-Ever-After with the One.'" In *Critical Thinking about Sex, Love, and Romance in the Mass Media: Media Literacy Applications*, edited by Mary-Lou Galician and Debra L. Merskin, 21–33. Mahwah, NJ: Lawrence Erlbaum Associates.

McClanahan, Andrea M. 2007. "Must Marry TV: The Role of Heterosexual Imaginary in the Bachelor." In *Critical Thinking about Sex, Love, and Romance in the Mass Media*, edited by Mary-Lou Galician and Debra L. Merskin, 261–274. Mahwah, NJ: Lawrence Erlbaum Associates.

McRobbie, Angela. 2004. "Post-Feminism and Popular Culture." *Feminist Media Studies* 4 (3): 255–264.

Melucci, Alberto. 1996. *The Playing Self: Person and Meaning in the Planetary Society*. Cambridge: Cambridge University Press.

Mendelman, Roni. 2013. "A Single Woman for Sale: The Sad Reality of the Dating Websites." *Onlife*. February 24. Accessed September 22, 2014. www.onlife.co.il/%D7%9E%D7%A9%D7%A4%D7%97%D7%94/%D7%99%D7%97%D7%A1%D7%99%D7%9D/52874/%D7%A8%D7%95%D7%95%D7%A7%D7%94-%D7%9C%D7%9E%D7%9B%D7%99%D7%A8%D7%94-%D7%94%D7%9E%D7%A6%D7%99%D7%90%D7%95%D7%AA-%D7%94%D7%A2%D7%92%D7%95%D7%9E%D7%94-%D7%A9%D7%9C-%D7%90%D7%AA%D7%A8%D7%99-%D7%94%D7%94%D7%99%D7%9B%D7%A8%D7%95%D7%99%D7%95%D7%AA.

Michael, Peter. 2014. "Man Drought Sees Shortage of Eligible Men as Women Struggle in Dating Game." *Courier Mail*. January 19. Accessed June 20, 2014. www.couriermail.com.au/news/queensland/man-drought-sees-shortage-of-eligible-men-as-women-struggle-in-dating-game/story-fnihsrf2-1226804921893.

Mohanty, Chandra T., Ann Russo, and Lourdes Torres. 1991. *Third World Women and the Politics of Feminism*. Bloomington, IN: Indiana University Press.

Montemurro, Beth. 2003. "Sex Symbols: The Bachelorette Party as a Window to Change in Women's Sexual Expression." *Sexuality and Culture* 7 (2): 3–29.

———. 2006. *Something Old, Something Bold: Bridal Showers and Bachelorette Parties*. New Brunswick, NJ: Rutgers University Press.

Moore, Jennifer A. and H. Lorraine Radtke. 2015. Starting "Real" Life: Women Negotiating a Successful Midlife Single Identity. *Psychology of Women Quarterly*, 1–15.

Moore, Wilbert E. 1963. *Man, Time and Society*. New York: John Wiley and Sons.

Moran, Rachel F. 2004. "How Second-Wave Feminism Forgot the Single Woman." *Hofstra Law Review* 33 (1): 223–298.

Morgan, David. 1996. *Family Connections: An Introduction to Family Studies*. Cambridge: Polity Press.

———. 2011. *Rethinking Family Practices*. Basingstoke: Palgrave Macmillan.

Morrow, Lance. "Waiting as a Way of Life." *Time*, July 23, 1984.

Nakano, Lynne. 2011. "Working and Waiting for an 'Appropriate Person': How Single Women Support and Resist Family." In *Home and Family in Japan: Continuity and Transformation*, edited by Richard Ronald and Allison Alexy, 51–131. London: Routledge.

———. 2014. "Single Women in Marriage and Employment Markets in Japan." In *Capturing Contemporary Japan: Differentiation and Uncertainty*, edited by Satsuki Kawano, Glenda S. Roberts and Susan O. Long, 163–182. Honolulu: University of Hawaii Press.

Needham, Gary. 2008. "Scheduling Normativity: Television, the Family and Queer Temporality." In *Queer TV: Theories, Histories, Politics*, edited by Glyn Davis and Gary Needham, 143–158. Abingdon, Oxon: Routledge.

Negra, Diane. 2009. *What a Girl Wants? Fantasizing the Reclamation of Self in Postfeminism*. Abington, Oxon: Routledge.

Netz, Tali. 2008. "I Wonder Where I Will Find Him: It Is Only a Matter of Time." *Ynet*. November 18. Accessed March 25, 2012. www.ynet.co.il/articles/0,7340,L-3624969,00.html.

Nowotny, Helga. 1992. "Time and Social Theory: Towards a Social Theory of Time." *Time & Society* 1 (3): 421–454.

O'Connor, Pat. 1993. "Same-Gender and Cross-Gender Friendships among the Frail Elderly." *The Gerontologist* 33 (1): 24–30.

———. 1998. "Women's Friendships in a Post-Modern World." In *Placing Friendship in Context*, edited by Rebecca G. Adams and Graham Allan, 117–135. Cambridge: Cambridge University Press.

Odeta. 2004. "How to Date Most Successfully." *nrg*. September 24. Accessed March 5, 2007. www.nrg.co.il/online/1/ART/786/844.html.

Odih, Pamela. 1999. "Gendered Time in the Age of Deconstruction." *Time & Society* 8 (1): 9–38.

———. 2007. *Gender and Work in Capitalist Economies*. Maidenhead, England: Open University Press and McGraw-Hill Education.

Office for National Statistics. 2015. "Population Estimates by Marital Status and Living Arrangements—England and Wales, 2002 to 2014." July 8. Accessed August 24, 2015. www.ons.gov.uk/ons/dcp171778_409686.pdf.

Oleksy, Elżbieta H. 2011. "Intesectionality at the Cross-Roads." *Women's Studies International Forum* 34 (4): 263–270.

Or, Ofra. 2013. "Midlife Women in Second Partnerships Choosing Living Apart Together: An Israeli Case Study." *Israel Studies Review* 28 (2): 41–60.

Or, Ori. 2011. "I Am Thirty Years Old and I Am Afraid of the Thin Line Separating Compromising and Giving Up." *Ynet*. January 12. Accessed May 10, 2012. www.ynet.co.il/articles/1,7340,L-4011851,00.html.

Or-Li, Tamar. 2008. "You Don't Want at the Age of Forty-Five to be Told that it is Too Late." *Ynet*. May 18. Accessed March 5, 2011. www.ynet.co.il/articles/0,7340,L-3544341,00.html.

Page, Susan. 2002. *If I'm so Wonderful Why Am I Still Single?* New York: Three Rivers Press.

Patton, Susan. 2013. "Letter to the Editor: Advice for the Young Women of Princeton: The Daughters I Never Had." *Daily Princetonian*. March 29. Accessed May 10, 2013. http://dailyprincetonian.com/opinion/2013/03/letter-to-the-editor-advice-for-the-young-women-of-princeton-the-daughters-i-never-had/.

———. 2014. *Marry Smart: Advice for Finding THE ONE*. New York: Gallery Books.

Pe'er, Ella. 2007. "I Am Thirty-One and No Man Has Ever Loved Me." *Ynet*. May 15. Accessed April 4, 2013 www.ynet.co.il/articles/0,7340,L-3399946,00.html.

Portuguese, Jacqueline. 1998. *Fertility Policy in Israel: The Politics of Religion, Gender, and Nation*. Westport, CT: Praeger Publishers.

Povoledo, Elisabetta. 2015. "Raw Eggs and No Husband Since '38 Keep Her Young at 115." *New York Times*. February 14. Accessed February 20, 2015. www.nytimes.com/2015/02/15/world/raw-eggs-and-no-husband-since-38-keep-her-young-at-115.html?_r=1.

Preser, Ruth. 2011. "Coherent Deviants: Transformation and Transition in Life-Stories of Once-Married Women Who Chose to Live as Lesbians." *Women's Studies International Forum* 34 (2): 140–150.

Press, Andrea and Sonia Livingstone. 2006. "Taking Audience Research into the Age of New Media: Old Problems and New Challenges." In *The Question of Method in Cultural Studies*, edited by Mimi White and James Schwoch, 175–200. Oxford: Blackwell Publishing.

Quirkyalone. 2015. "About." Accessed December 25, 2015. http://quirkyalone.net/index.php/about-2/.

Regev, Yael. 2010. "The Picky Woman." *Ynet*. April 3. Accessed November 2, 2013. www.ynet.co.il/articles/0,7340,L-3870823,00.html.

Reinstein, Ziv. 2009. "They Are Celebrating One Year and Now They Have a Baby on the Way." *Ynet*. February 1. Accessed March 15, 2010. www.ynet.co.il/articles/0,7340,L-3664588,00.html.

Reith, Gerda. 1999. "In Search of Lost Time: Recall, Projection and the Phenomenology of Addiction." *Time and Society* 8 (1): 99–117.

Remennick, Larissa. 2000. "Childless in the Land of Imperative Motherhood: Stigma and Coping among Infertile Israeli Women." *Sex Roles* 43 (11–12): 821–841.

Resnik, Merav. 2006. "The Heart is Bleeding, Yet it Refuses to Evacuate from the Danger Zone." *Ynet*. November 29. Accessed April 17, 2012. www.ynet.co.il/articles/0,7340,L-3333546,00.html#n.

———. 2007a. "You Freeze in Fear and then You Miss the Train." *Ynet*. May 2. Accessed July 23, 2013. www.ynet.co.il/articles/0,7340,L-3394797,00.html.

———. 2007b. "Not Beautiful? Go to Look for Another Dating Website." *Ynet*. October 1. Accessed November 12, 2011. www.ynet.co.il/articles/0,7340,L-3454443,00.html.

———. 2007c. "How is He Still Single? Maybe There is Something Wrong with Him." *Ynet*. March 7. Accessed April 9, 2014. www.ynet.co.il/articles/0,7340,L-3372668,00.html.

Reynolds, Jill. 2008. *The Single Woman: A Discursive Investigation*. London: Routledge.

Reynolds, Jill and Margaret Wetherell. 2003. "The Discursive Climate of Singleness: The Consequences for Women's Negotiation of a Single Identity." *Feminism and Psychology* (13): 489–510.

Rhodin, Sara. 2008. "A Holiday from Russia with Love." *New York Times*. July 9. Accessed March 12, 2014. www.nytimes.com/2008/07/09/world/europe/09russia.html.

Rich, Adrienne. 1980. "Compulsory Heterosexuality and Lesbian Existence." *Signs* 5 (4): 631–660.

Rivers, Caryl. 2006. "Newsweek's Apology Too Little, 20 Years Too Late." *AlterNet*. June 14. Accessed February 21, 2012. www.alternet.org/story/37582/newsweek's_apology_too_little,_20_years_too_late.

Ronai, Carol R. 2000. "Managing Aging in Young Adulthood: The 'Aging' Table Dancer." In *Aging and Everyday Life*, edited by James A. Holstein and Jaber F. Gubrium, 277–287. Oxford: Blackwell Publishing.

Roseneil, Sasha. 2004. "Why We Should Care about Friends: An Argument for Queering the Care Imaginary in Social Policy." *Social Policy and Society* 3 (4): 409–419.

Roseneil, Sasha and Shelley Budgeon. 2004. "Cultures of Intimacy and Care Beyond 'the Family': Personal Life and Social Change in the Early 21st Century." *Current Sociology* 52 (2): 135–159.

Roth, Julius A. 1963. *Timetables: Structuring the Passage of Time in Hospital Treatment and Other Careers*. Indianapolis: Bobbs-Merrill.

Rothman, Barbara K. 2005. *Weaving a Family: Untangling Race and Adoption*. Beacon Press.

Sa'ar, Dana. 2007. "A Break? Who Needs a Timeout from Sex?" *Ynet*. May 9. Accessed March 20, 2010. www.ynet.co.il/articles/0,7340,L-3397258,00.html.

Sa'ar, Tsafi. 2012. "If There Are Parties without Women, Why Wouldn't There Be Parties without Mizrachi Jewish?" *Haaretz*. February 14. Accessed July 6, 2015. www.haaretz.co.il/gallery/mejunderet/1.1641008.

Salholz, Eloise. 1986. "The Marriage Crunch." *Newsweek*. June 2.

Sandfield, Anna and Carol Percy. 2003. "Accounting for Single Status: Heterosexism and Ageism in Heterosexual Women's Talk about Marriage." *Feminism & Psychology* 13 (4): 475–488.

Scarce, Rik. 2002. "Doing Time as an Act of Survival." *Symbolic Interaction* 25 (3): 303–321.

Schefft, Jen. 2007. *Better Single than Sorry: A No Regrets Guide to Loving Yourself and Never Settling*. New York: HarperCollins Publishers.

Schippers, Mimi. 2007. Recovering the Feminine Other: Masculinity, Femininity, and Gender Hegemony. *Theory and Society* 36: 85–102.

Schubert, Violeta D. 2009. "Out of 'Turn', Out of Sync: Waiting for Marriage in Macedonia." In *Waiting*, edited by Ghassan Hage, 107–120. Melbourne, VIC: Melbourne University Press.

Schwartz, Barry. 1975. *Queuing and Waiting: Studies in the Social Organization of Access and Delay*. Chicago: University of Chicago Press.

Schwartzberg, Natalie, Kathy Berliner, and Demaris Jacob. 1995. *Single in A Married World: A Life Cycle Framework for Working with the Unmarried Adult*. New York: W. W. Norton & Company.

Sela, Li-Or and Adi Kimchi. 2011. "*Bekarov Ezlech*: The Single Woman and the Pressure which Accompanies Her." *Ynet*. December 18. Accessed May 15, 2013. www.ynet.co.il/articles/0,7340,L-4163350,00.html.

Shalev, Carmel and Sigal Gooldin. 2006. "The Uses and Misuses of in Vitro Fertilization in Israel: Some Sociological and Ethical Considerations." *Nashim: A Journal of Jewish Women's Studies & Gender Issues* 12 (1): 151–176.

Shalom, Moriah. 2006. "When Searching Becomes Your Main Occupation." *Ynet*. October 15. Accessed August 18, 2009. www.ynet.co.il/articles/0,7340,L-3315546,00.html.

Shamir, Michal. 2012. "Team Building Day at Work: Everyone is with Spouses, I'm the Only One Alone." *Ynet*. September 5. Accessed March 20, 2013. www.ynet.co.il/articles/0,7340,L-4277624,00.html.

Shargal, Dvorit. 2006. "Alone." *Ynet*. August 2. Accessed June 25, 2013. www.ynet.co.il/articles/0,7340,L-3285474,00.html.

Sharp, Elizabeth A. and Lawrence Ganong. 2011. "'I'm A Loser, I'm Not Married, Let's just All Look at Me': Ever-Single Women's Perceptions of their Social Environment." *Journal of Family Issues* 32 (7): 956–980.

Shechter, Asher. 2015. *The Marker*. "I'm 43 Years Old Now, and My Spinster Wish Has Been Fulfilled." September 17. Accessed October 2, 2015. www.themarker.com/magazine/1.2715993#hero__bottom.

Shweder, Richard A. (Ed.) 1998. *Introduction to Welcome to Middle Age! (and Other Cultural Fictions)*, ix–xvii. Chicago: University of Chicago Press.

Silver, Catherine B. 2003. "Gendered Identities in Old Age: Toward (De)Gendering?" *Journal of Aging Studies* 17 (4): 379–397.

Simmel, Georg. 1997. "The Adventure." In *Simmel on Culture: Selected Writings*, edited by Georg Simmel, David Frisby, and Mike Featherstone, 221–232. London: Sage Publications.

Simpson, Roona. 2003. "Contemporary Spinsters in the New Millennium: Changing Notions of Family and Kinship." New Working Paper Series 10. London: Gender Institute, London School of Economics and Political Science. Accessed September 5, 2010. http://eprints.lse.ac.uk/37936/.

———. 2006. "The Intimate Relationships of Contemporary Spinsters." *Sociological Research Online* 11 (3).

Single Edition. 2008. "Home." Accessed January 15. www.singleedition.com/.

Song, Jesook. 2010. "'A Room of One's Own': The Meaning of Spatial Autonomy for Unmarried Women in Neoliberal South Korea." *Gender, Place and Culture* 17 (2): 131–149.

———. 2014. *Living on Your Own: Single Women, Rental Housing, and Post-Revolutionary Affect in Contemporary South Korea*. Albany: State University of New York Press.

Sontag, Susan. 1983. "The Double Standard of Aging." In *On the Contrary: Essays by Men and Women*, edited by Martha Rainbolt and Janet Fleetwood, 99–112. Albany: State University of New York Press.

Spector, Dana. 2004. "A Guide for the Desperate Single Woman: How to Market Yourself for the Man of Your Dreams." *Ynet*. June 25. Accessed January 15, 2012. www.ynet.co.il/articles/0,7340,L-2937167,00.html.

Speed-Date. 2014. "What is a Speed-Date?" Accessed July 5, 2014. www.speed-date.co.il/whatis.aspx.

Spelman E.V. 1982. "Woman as Body: Ancient and Contemporary Views." *Feminist Studies*, 8 (1): 109–131.

Spencer, Liz and Raymond Edward Pahl. 2006. *Rethinking Friendship: Hidden Solidarities Today*. Princeton, NJ: Princeton University Press.

Statistics Denmark. 2016. "FAM44N: Families 1. January by Municipality, Type of Family, Size of Family and Number of Children." Accessed August 15, 2016. www.statbank.dk/statbank5a/SelectVarVal/Define.asp.

Stein, Karen. 2012. "Time Off: The Social Experience of Time on Vacation." *Qualitative Sociology* 35 (3): 335–353.

Stern, Rona. 2007. "Don't You Think I Am Making a Pass at You." *Ynet*. February 27. Accessed June 6, 2009. www.ynet.co.il/articles/0,7340,L-3369454,00.html.

Stewart, Jerusha. 2005. *The Single Girl's Manifesta: Living in a Stupendously Superior Single State of Mind*. Naperville, IL: Sourcebooks Casablanca.

Stromza, Sivan. 2012. "Old Maid." *Stromz* (blog), November 28. http://saloona.co.il/stromz/?p=144.

Sudnow, David. 1967. *Passing On: The Social Organization of Dying*. Englewood Cliffs, NJ: Prentice Hall.

Swidler, Ann. 2003. *Talk of Love: How Culture Matters*. Chicago: University Of Chicago Press.

Tal-Meir, Kinneret. 2013. "Singles, this Holiday You Will Beat the Pitying Looks." *Ynet*. September 4. Accessed January 2, 2014. www.ynet.co.il/articles/0,7340,L-4425522,00.html.

Taylor, Anthea. 2011. "Blogging Solo: New Media, 'Old' Politics." *Feminist Review* 99: 79–97.

———. 2012. *Single Women in Popular Culture: The Limits of Postfeminism*. New York: Palgrave Macmillan.

Teman, E. (2010). *Birthing a Mother: The Surrogate Body and the Pregnant Self*. Berkeley, University of California Press.

The Naked Truth. 2008. "Why Am I Afraid of Going Out to a Bar Alone?" *Ynet*. March 21. Accessed August 3, 2009. www.ynet.co.il/articles/0,7340,L-3521873,00.html.

Thelma and Louise. 2006a. "You Shall Rejoice and Lose Your Head." *Ynet*. September 22. Accessed July 6, 2014. www.ynet.co.il/articles/0,7340,L-3306976,00.html.

———. 2006b. "Thus the Typical Single Woman Saved Next Year." *Ynet*. December 29. Accessed March 2, 2014. www.ynet.co.il/articles/0,7340,L-3345917,00.html.

Thompson, Edward P. 1967. "Time, Work-Discipline, and Industrial Capitalism." *Past and Present* 38: 56–97.

Torre, Ramón R. 2007. "Time's Social Metaphors: An Empirical Research." *Time & Society* 16 (2–3): 157–187.

Traister, Rebecca. 2016a. *All the Single Ladies: Unmarried Women and the Rise of an Independent Nation*. New York: Simon and Schuster.

Traister, Rebecca. 2016b. "The Single American Woman." *New York*. February 22. Accessed June 11, 2016. http://nymag.com/thecut/2016/02/political-power-single-women-c-v-r.html.

Trimberger, Ellen K. 2005. *The New Single Woman*. Boston, MA: Beacon Press.

Turkle, Sherry. 1997. *Life on the Screen: Identity in the Age of the Internet*. New York: Touchstone.

Turner, Bryan S. and Steven P. Wainwright. 2003. "Corps De Ballet: The Case of the Injured Ballet Dancer." *Sociology of Health & Illness* 25 (4): 269–288.

Turner, Victor. 1969. *The Ritual Process: Structure and Anti-Structure*. Ithaca, NY: Cornell University Press.

Tye, Diane and Ann M. Powers. 1998. "Gender, Resistance and Play: Bachelorette Parties in Atlantic Canada." *Women's Studies International Forum* 21 (5): 551–561.

United States Census Bureau 2016. "Adults (A table series)—People and Households." Accessed July 26, 2016. www.statistikbanken.dk/Statbank5a/SelectVarVal/Define.asp?MainTable=FAM44N&PLanguage=1&PXSId=0&wsid=cftree.

Unmarried America. 2015. "National USA [Unmarried and Single Americans] Week Home." Accessed November 28, 2015. www.unmarriedamerica.org/usaweek/intro.htm.

Van Dijk, Teun. 1993. "Principles of Critical Discourse Analysis." *Discourse and Society* 4 (2): 249–283.

Van Gennep, Arnold. 1960. *The Rites of Passage*. Chicago: University of Chicago Press.

Van Hooff, Jenny. 2013. *Modern Couples? Continuity and Change in Heterosexual Relationships*. Farnham: Ashgate Publishing Limited.

Vint, Sherryl. 2007. "The New Backlash: Popular Culture's 'Marriage' with Feminism, or Love is all You Need." *Journal of Popular Film and Television* 34 (4): 160–169.

Vitus, Kathrine. 2010. "Waiting Time: The De-Subjectification of Children in Danish Asylum Centres." *Childhood* 17 (1): 26–42.

Waldby, Catherine. 2015. "'Banking Time': Egg Freezing and the Negotiation of Future Fertility." *Culture, Health & Sexuality* 17 (4): 470–482.

Wallace, Kelly. 2014. "Should You Go to College for Mrs. Degree? Princeton Mom Weighs In." *CNN*. March 14. Accessed April 7, 2014. http://edition.cnn.com/2014/03/13/living/princeton-mom-book-marry-smart-matrimony/.

Warner, Michael. 1993. *Introduction to Fear of a Queer Planet: Queer Politics and Social Theory*, edited by Michael Warner, vii–xxxi. Minneapolis: University of Minnesota Press.

———. 1999. *The Trouble with Normal: Sex, Politics, and the Ethics of Queer Life*. New York: Free Press.

Weber, Max. 1985. *The Protestant Ethic and the Spirit of Capitalism*. New York: Scribner's.

Weston, K. 1991. *Families We Choose: Gay and Lesbian Kinship*. New York: Columbia University Press.

Whitelocks, Sadie. 2014. "'Girls Should Get Major Bodywork in High School': Controversial Princeton Mom Says Cosmetic Surgery Will Help Women Find Love." *Daily Mail*. March 11. Accessed June 7, 2014. www.dailymail.co.uk/femail/article-2578499/Girls-major-bodywork-high-school-Controversial-Princeton-Mom-says-cosmetic-surgery-help-women-love.html.

Wilkinson, Eleanor. 2012. "The Romantic Imaginary: Compulsory Coupledom and Single Existence." In *Sexualities: Past Reflections, Future Directions*, edited by Sally Hines and Yvette Taylor. Basingstoke: Palgrave Macmillan.

———. 2014. "Single People's Geographies of Home: Intimacy and Friendship Beyond 'the Family.'" *Environment and Planning A* 46 (10): 2452–2468.

Woodward, Kathleen M. 2006. "Performing Age, Performing Gender." *NWSA Journal* 18 (1): 162–189.

Yael9. 2006. "The Single Woman's Speech: When Will You Bring a Doctor to the Family?" *Ynet*. April 13. Accessed June 6, 2008. www.ynet.co.il/articles/0,7340,L-3238200,00.html.

Yechimovich, Hila. 2013. "She is Forty Years Old and She is Still Waiting for the Knight on the White Horse." *Mako*. January 5. Accessed February 26, 2015. www.mako.co.il/women-sex_and_love/couples/Article-1aead222f1dcb31006.htm.

Yian, Hogne. 2004. "Time Out and Drop Out: On the Relation between Linear Time and Individualism." *Time & Society* 13: 173–195.

Yishi, Harela. 2013. "From the Queen of the Class to a Forty-Year-Old Single Woman: How Did It Happen?" *Ynet*. April 19. Accessed December 15, 2014. www.ynet.co.il/articles/0,7340,L-4369414,00.html.

Yle. 2014. "NGO: Singles Facing High Costs, Discrimination." June 7. Accessed September 18, 2014. http://yle.fi/uutiset/ngo_singles_facing_high_costs_discrimination/7284504.

Young, Michael D. and Tom Schuller. 1991. *Life after Work: The Arrival of the Ageless Society*. London: HarperCollins Publishers.

Zerubavel, E. 1981. *Hidden Rhythms: Schedules and Calendars in Social Life*. Chicago: University of Chicago Press.

———. 1985. *The Seven Day Circle*. New York: Free Press Macmillan.

Additional resources

27 Dresses. Directed by Anne Fletcher. New York: Fox 2000 Pictures, Spyglass Entertainment and Dune Entertainment, 2008.

Bridesmaids. Directed by Paul Feig. Los Angeles, CA: Apatow Productions and Relativity Media, 2011.

Elite-Strauss. "Sweetening it for Single Women." Saatchi & Saatchi, directed by Eran Shefnir, 2013.

Friends. "The One Where Chandler Crosses the Line." Episode No. 80, first broadcast November 13, 1997 by NBC. Directed by Kevin S. Bright and written by Adam Chase.

Mesudarim. Directed by Shachar Berlovitch and Oded Reskin. Created by Assaf Harel and Muli Segev. Keshet, 2007.

The Wedding Date. Directed by Clare Kilner. London and Shere, Surrey: Gold Circle Films, 26 Films and Visionview Production, 2005.

Index

EU authorised representative for GPSR:
Easy Access System Europe, Mustamäe tee 50,
10621 Tallinn, Estonia
gpsr.requests@easproject.com